STRATEGIES FOR SOLVING

MATH WORD PROBLEMS

by Jerome D. Kaplan, Ed. D.

Editorial Director: **Caleb E. Crowell**

EDUCATIONAL DESIGN, INC.   EDI 323

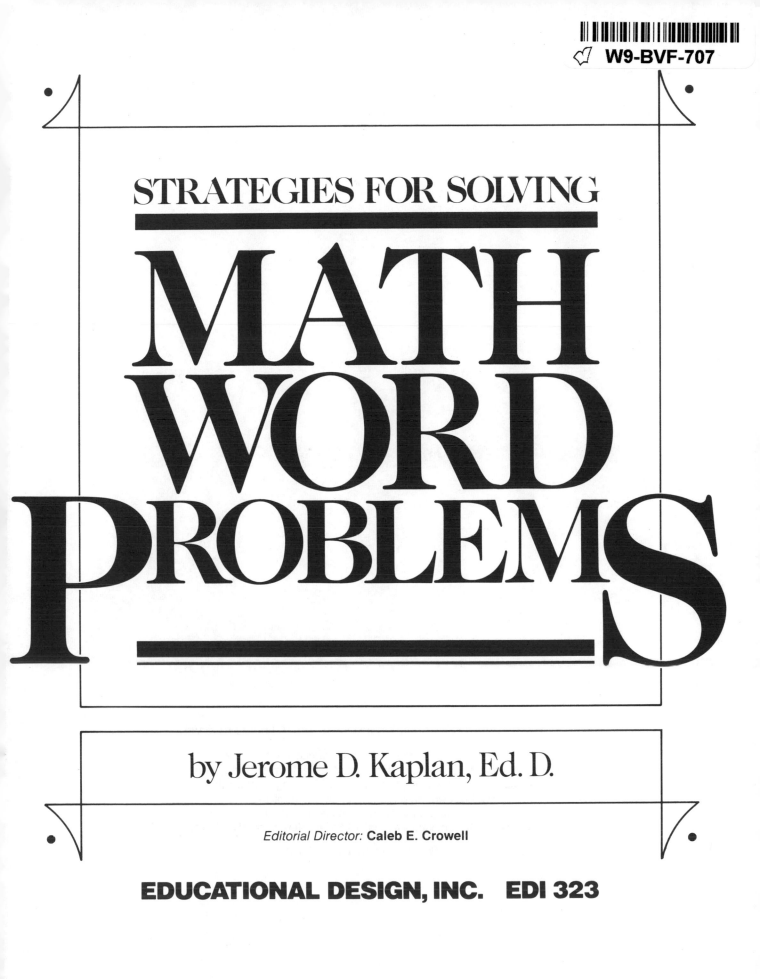

W9-BVF-707

ISBN# 0-87694-074-2   EDI 323

Table of Contents

1. How to Solve Word Problems

1.1 Introduction

Are you good at solving word problems? If you answered "no," you are not alone. Most students have trouble with word problems. For lots of students, they are the hardest part of math. Math teachers often talk about the difficulty their students have with word problems. For this reason we have come up with a series of strategies to help you solve word problems.

In this chapter we introduce the main strategy. It is a simple one—so simple that you may wonder what the big deal is. This strategy is the basis for all other strategies. You can apply it to all word problems in your math studies and books, not only this book.

The strategy follows common sense. It consists of five obvious steps. But even if these steps seem simple, we urge you to follow them carefully. Do not leave any step out. Too often people disregard a step because they think the steps are so obvious they need not follow them. We don't want that to happen to you. So when you solve a problem in this book we want you to always use all the steps of the strategy introduced in this chapter.

1.2 Five Steps for Solving Word Problems

In this section we will show you the key strategy in this book. The strategy consists of five steps for solving word problems. Then in the rest of this chapter we will apply these steps to a number of examples.

Here are the five steps:

WHAT   HOW   HELP   COMPUTE   CHECK

Here's what they mean:

1. WHAT?

Ask yourself: What does the problem want you to find? What is the question asking for? An area, or the number of people, or a cost, or an average, or the height of a building? In this step, you identify what the problem asks you to find. If you are careless with this step, your answer will be wrong.

2. HOW?

How will you do this problem? What mathematical operation must you use—do you add, or subtract, or multiply, or divide? Will you need to do more than one operation? In this step, you look at the information in the problem and decide what operation or operations will give you your answer. This step requires the most thinking of all.

3. HELP

You probably won't be able to do the math in your head. You will need help. Setting up your computation is one kind of help. Using a formula, or a graph, or a table, or drawing a diagram are other kinds of help.

4. COMPUTE

This is the step where you actually "do the math." At the end of this step, you should have your answer. This is the one step that nobody ever forgets to do!

5. CHECK

At the end of Step 4, you have your answer. But it may not be correct. Everybody makes mistakes. You have to check your answer to make sure it is right. There are several ways to check your answer that you will learn in this book. This step is the only way to catch the mistakes you make.

These are the five simple steps that you will need to solve problems throughout this book and elsewhere. Think about them for a minute. There is a logic to each step. Try to remember the five steps and the order they are in.

- -

You may find it useful to learn a strategy for remembering the five steps.

WHAT   HOW   HELP   COMPUTE   CHECK

Make a code out of them. Here's the code invented by some students to remember these five steps. They made up a sentence whose words contain the names of all of the steps, in order:

WHAT!! HOWARD HELPS COMPUTE CHECKS?

A sentence like this, which helps you remember something, is often called a memory gem. This particular memory gem uses all of the words of the five steps almost exactly.

To help you remember the memory gem, we offer you a simple story:

Lazy Howard was the laughing stock of the bank where he worked. He took long breaks, made errors counting money, and often spilled coffee on important computer printouts. Every afternoon around 3 o'clock he disappeared for about an hour. Everyone wondered where he went. One afternoon Wanda opened the door to the computer room. Was she surprised! There was Howard helping the treasurer of the bank compute payroll checks. She burst out:

"WHAT! HOWARD HELPS COMPUTE CHECKS?"

A silly story, you say, but we hope it helps you remember the five steps.

✎ *Exercises*

Circle the correct answer in each of these exercises.

1. Susan decides to use multiplication to solve a word problem. Which step is this?

> WHAT   HOW   HELP   COMPUTE   CHECK

2. Larry decides that a word problem wants him to find the distance between two towns. Which step is this?

> WHAT   HOW   HELP   COMPUTE   CHECK

3. Marcia looks over her answer to a word problem and discovers it was wrong. Which step is this?

> WHAT   HOW   HELP   COMPUTE   CHECK

4. Leroy knows he must divide the number 216 by 12 to find the answer to a word problem. He writes this:

$$12\overline{)216}$$

Which step is this?

> WHAT   HOW   HELP   COMPUTE   CHECK

5. Julia is working on a word problem. As part of her work, she subtracts 216 from 384 and gets 168. Which step is this?

> WHAT   HOW   HELP   COMPUTE   CHECK

6. Martin thinks about his answer to a word problem that asked him to find the profit on a store's sale of an inexpensive radio. His answer was $14,730. Martin decides his answer doesn't seem reasonable. Which step is this?

> WHAT   HOW   HELP   COMPUTE   CHECK

7. Alicia is trying to figure out the area of a field. The problem contains a formula that she can use. The formula is an example of—

> WHAT   HOW   HELP   COMPUTE   CHECK

1.3 Solving a Problem—An Example

Throughout the book you will be solving word problems from many areas of mathematics—problems with whole numbers, fractions, statistics, measurement, and other topics.

We want you to use the five-step strategy for all these kinds of problems. So in this section we are going to demonstrate each step, using a single example. We want you to understand the five steps very well, so that when you solve problems in other chapters and elsewhere, you will know what to do.

Here is the example:

EXAMPLE

Joan opened a checking account with a deposit of $345.89 on Monday. On Wednesday, she withdrew $79.98. How much money does she have left in her account?

Read this problem as often as you want until you understand it. Go back and read it again whenever you need to do so. We will soon solve this problem, using the five-step strategy introduced in the previous section.

Do you remember the memory gem sentence that some students made up to help them remember the five steps? Here it is:

WHAT! HOWARD HELPS COMPUTE CHECKS?

Now that you know this strange sentence, do you remember the five steps? In the next sections of this chapter, we shall go through each step for the example above.

1. What step does each word of the memory gem sentence above stand for? Fill in the name of the step. (For a couple of steps, all you have to do is copy the word.)

   a. WHAT   stands for _____.
   b. HOWARD   stands for _____.
   c. HELPS   stands for _____.
   d. COMPUTE   stands for _____.
   e. CHECKS   stands for _____.

2. In the example at the beginning of this section, which of these facts is important in finding the answer?

   a. The name of the person is <u>Joan</u>.
   b. Joan deposited money on <u>Monday</u>.
   c. Joan deposited <u>$345.89</u> in her account.
   d. Joan withdrew money on <u>Wednesday</u>.

3. Which of these means the same as the words "a <u>deposit</u> of $345.89" in the example?

   a. amount Joan earns at her job every week
   b. amount Joan places into the account
   c. amount Joan takes out of the account
   d. amount Joan owes the bank

4. Which of these words means the same as "withdrew" in the example?

   a. bought
   b. left in
   c. put in
   d. took out

1.4 The First Step—WHAT?

The first step of the five-step strategy asks WHAT?

- WHAT does the problem want you to answer?

- WHAT type of answer are you looking for? Is it the length of a river, the cost of a house, the amount of money earned, or the distance traveled, or something else?

The WHAT? step gets you started thinking about the whole problem. When you answer the question WHAT? you have a good idea of where you will be heading.

The WHAT? part of a word problem is always easy to find. It is in the sentence that ends with a question mark. That's usually the last sentence of the problem. To find out WHAT, just zero in on this question sentence.

Here is the same example of the last section. Study this example to see how the first step gets you off on the right course. Zero in on the sentence with the question mark—the question sentence.

EXAMPLE

Joan opened a checking account with a deposit of $345.89 on Monday. On Wednesday, she withdrew $79.98. How much money does she have left in her account?

✎ Exercises

Answer these questions about the example. Choose the best answer.

1. WHAT question does the problem want answered? (Look at the question sentence.)

   a. How much money did Joan withdraw from her account?
   b. How much money did Joan deposit in her account?
   c. How much money was left in Joan's account?

2. WHAT will the answer be?

   a. The amount of money in the account after Joan withdrew $79.98
   b. The amount of money at the start of the account
   c. The amount of money withdrawn from the account

3. Which of these questions does the WHAT? step ask?

   a. WHAT does the problem want you to answer?
   b. WHAT is the answer that you have calculated?
   c. WHAT operation will solve this problem?

1.5 The Second Step—HOW?

The second step of the five-step strategy asks HOW?

- HOW is the problem solved?

- HOW are "key words" used to guide you in choosing the correct operation? (You'll find out about "key words" later. They are part of a special strategy for the HOW? step.)

- HOW are mathematical operations used to solve the problem?

The HOW? step gets you into the details, facts, and numbers given in the problem. After you answer the HOW? step, you will be ready to start figuring out the solution to the problem.

The HOW? step is the step that takes the most real thinking. It is really the heart of solving a word problem. Later in this chapter you will be given some strategies for finding out HOW. But the basic strategy is to read the details of the problem carefully with the WHAT? question in mind.

Here is the same example as in the last section. Study this example to see how the HOW? step gets you into the actual figuring out of the solution.

EXAMPLE

Joan opened a checking account with a deposit of $345.89 on Monday. On Wednesday, she withdrew $79.98. How much money does she have left in her account?

✎ *Exercises*

These questions are about the example. Choose the correct answer for each one.

1. Which of these underlined words gives you the best hint on HOW to solve the problems?

    a. Joan <u>opened</u> a checking account.
    b. Joan made a <u>deposit</u>.
    c. How much money does she have <u>left</u> in her account?

2. Which operation do you use to solve this problem?

    a. addition
    b. subtraction
    c. division

3. Which of these questions does the HOW? step ask?

    a. How can the problem be reworded?
    b. How can the answer be checked?
    c. How can the problem be solved?

1.6 The Third Step—HELP

The third step of the five-step strategy tells you to ask for HELP.

- Do you need HELP? (You do if you can't do all the math in your head.)

- Do you need extra HELP from a picture, a diagram or an equation?

- With every problem, there is a computation setup that can HELP. Do it neatly and carefully.

The HELP step asks if you need HELP, or if you can get some HELP from somewhere. Even a simple diagram or an equation might HELP. A computation setup is almost always necessary.

Sometimes you may be in a hurry. You may think that it's a bother to slow down and set up your computation neatly and carefully. But if you set up your computation carelessly, you may not do the math correctly.

Again we look at the same example that we've been using for illustration. What kind of HELP do you need for this example?

EXAMPLE

Joan opened a checking account with a deposit of $345.89 on Monday. On Wednesday, she withdrew $78.98. How much money does she have left in her account?

✎ *Exercises*

Answer these questions about the example above. Choose the correct answer for each question.

1. What kind of HELP would be best for solving this example?

    a. a diagram of a bank
    b. a table of interest rates
    c. a computation setup

2. Which of these computation setups will be the best HELP in solving the example?

    a. $345.89 - 78.98$

    b.
$$\begin{array}{r} 345.89 \\ -\ 78.98 \\ \hline \end{array}$$

    c.
$$\begin{array}{r} 345.89 \\ -\ 78.98 \\ \hline \end{array}$$

3. Which of these questions might you ask as part of the HELP step?

    a. Is the answer reasonable?
    b. Can a chart help solve the example?
    c. What question does the problem ask?

1.7 The Fourth Step—COMPUTE

The fourth step of the five-step strategy tells you to do the math. This is what COMPUTE means.

Here are some questions to think about during this step:

- What numbers are you using to COMPUTE?
- Do you have the right setup before you COMPUTE?
- Are you being careful as you COMPUTE?

The COMPUTE step is the step that gets you the answer. It is the step that you have been working towards. The first three steps before COMPUTE prepare you for this critical step.

Here is the same example as before.

EXAMPLE
Joan opened a checking account with a deposit of $345.89 on Monday. On Wednesday, she withdrew $78.98. How much money does she have left in her account?

✎ Exercises

Answer these questions about the example above. Choose the correct answer for each question.

1. What operation do you use in the COMPUTE step?

   a. addition
   b. subtraction
   c. multiplication
   d. division

2. COMPUTE the answer to the example above. The answer is

   a. 265.91
   b. 266.91
   c. 267.91

3. One of the questions that the COMPUTE step asks is—

   a. Which operation should you use?
   b. Which checking process do you use?
   c. Do you have the right setup before you COMPUTE?

1.8 The Fifth Step—CHECK

The fifth step of the five-step strategy is CHECK. Here are two questions to ask during this step:

- Does the answer make sense?
- Have you actually checked the answer?

The CHECK step is the last step. It is the step that lets you check the answer you got from the first four steps. If you made a computation to get the answer, then the last step checks to see if the answer is correct.

There are really two parts to the CHECK step:

Part 1 is to look at your answer and to use your common sense to see if it's reasonable. If, for example, your answer to the example was $2669.10, you would know that couldn't be right. Joan only deposited about $345. She couldn't possibly have more than $2000 left after taking money out of her account. You would know right away that you have to do the problem over.

Part 2 is a more exact kind of check. Do it after you are satisfied with Part 1. Starting with your answer, you do the opposite of the computation you performed in the COMPUTE step. If you subtracted, you would check by adding your answer to the numbers in your computation. You should end up with the original number if your answer to the COMPUTE step was correct.

Look at the example that we have been using for illustration.

EXAMPLE
Joan opened a checking account with a deposit of $345.89 on Monday. On Wednesday, she took out $78.98. How much money does she have left in her account?

The answer to this example is $266.91 (see Section 1.7). Answer the following questions by choosing the correct answer for each one.

1. Why is the answer $266.91 a <u>reasonable</u> answer?

    a. It is less than the original amount deposited in the bank.
    b. It is more than the original amount deposited in the bank.
    c. It is equal to the amount deposited in the bank.

2. How else can we check the computation for this example?

    a. addition
    b. subtraction
    c. multiplication
    d. division

3. Here's the COMPUTE step of the problem, and the Part 2 check that goes with it. Complete the check.

COMPUTE STEP	CHECK
345.89	266.91
−78.98	+78.98
266.91	

Is your answer the same as the first number in the problem? It should be.

4. Which one of these questions does the CHECK step ask?

    a. Is the answer reasonable?
    b. Is the answer a large number?
    c. Is the answer a small number?

1.9 Key Words

As you have seen by working through several problems in this section, there are certain words that offer clues to solving word problems. We call these words *key words*, and you will very often find them in the sentence that ends with a question mark—the sentence that you first focus on in the WHAT? step.

You use these key words, however, in the HOW step. They guide you in choosing the correct computation method for solving the problem. Using key words is a special strategy you should master.

In the list below, we have collected these key words for you. Try to memorize them and to remember what they mean so that you can use them when you solve problems. This list includes words from this chapter and words from later chapters. There aren't many of them, so take advantage of the few that there are.

You will find each key word followed by the operation that it suggests. Then you will see an example that uses each key word. This list will help you throughout the book. You should refer back to this list whenever you need to.

KEY WORDS

WORDS	OPERATION	EXAMPLE
altogether **total**	add <u>or</u> multiply	**(add)** Jeff picked up 43 marbles, 73 sea shells and 105 pins. How many things did he pick up **altogether?** (or, what is the **total** number of things he picked up? **(multiply)** Fran bought seven lottery tickets every day for five days. How many tickets did she buy *altogether*? (or, what is the **total** number of tickets she bought?)
left **remaining**	subtract	How much money is **left**? How much money is **remaining**?
more	subtract	How many **more** people live in New Jersey than in Colorado?
increase **go up** **grow**	subtract	How much did the price **increase** from last year? NOTE: This key word may surprise you. You might think that "increase" would mean to add. But it doesn't. To find an increase, you have to *subtract* the old, lower size or amount from the new, higher amount. The same goes for "go up" and "grow."
decrease **go down** **reduce**	subtract	How much did the price **decrease** from last year? NOTE: You may think that if "increase" means subtraction, then "decrease" would mean addition. But it doesn't. To find a decrease, you use exactly the same *subtraction* operation as for increase. The same goes for "go down" and "reduce."

. . . . *continued*

KEY WORDS

WORDS	OPERATION	EXAMPLE
of (particularly when used with fractions and percents)	multiply	**(fraction)** How many miles is 3/5 **of** the distance? **(percent)** What is 43% **of** the original $600?
each	divide	How many did **each** one get?
per	divide	What is the price **per** gallon?

✎ Exercises

Mark the correct answer for each question with a check mark.

1. Which key word suggests the operation **add?**

    a. left
    b. altogether
    c. each
    d. of

2. Which key word suggests the operation **subtract?**

    a. left
    b. altogether
    c. each
    d. of

3. Which key word suggests the operation **divide?**

    a. left
    b. altogether
    c. each
    d. of

4. Which key word suggests the operation **multiply?**

    a. left
    b. altogether
    c. each
    d. of

5. What operation is suggested by this question: How many <u>more</u> does he have to do?

    a. addition
    b. subtraction
    c. multiplication
    d. division

6. What operation is suggested by this question: How much did <u>each</u> one cost?

    a. addition
    b. subtraction
    c. multiplication
    d. division

7. What operation is suggested by this statement and question: It took $\frac{3}{4}$ of the summer vacation to build the addition to the kitchen. How much time was that?

    a. addition
    b. subtraction
    c. multiplication
    d. division

8. What operation is suggested by this statement and question: Jim worked 13 hours every day for 5 straight days. How many hours was that <u>altogether</u>? (Be careful—remember that this key word can suggest two different operations. You have to read the whole problem to find out which one goes here.)

    a. addition
    b. subtraction
    c. multiplication
    d. division

9. What operation is suggested by this question: How much was Laura's salary <u>increase</u>?

    a. addition
    b. subtraction
    c. multiplication
    d. division

10. What operation is suggested by this statement and question: Sixty percent <u>of</u> the senior class did not come to the first dance. How many people came to the dance? (Notice that the key word in this example is not in the sentence.)

    a. addition
    b. subtraction
    c. multiplication
    d. division

11. What operation is suggested by this statement and question: The team <u>gained</u> 75 yards less by rushing this week than it did last week. The team gained 315 yards last week. How much did the team <u>gain</u> this week?

    a. addition
    b. subtraction
    c. multiplication
    d. division

1.10 Using the Five Steps—I

In this section we apply the five steps to another problem.

EXAMPLE

> Jim bought 15 gallons of gasoline at the ABC Gas Station. He paid $16.50. What was the price per gallon?

We solve a problem with the five-step strategy. To remember the five steps we use the memory gem:

WHAT! HOWARD HELPS COMPUTE CHECKS?

1. WHAT—WHAT does the problem want us to find?

Examine the question part of the problem. The question sentence asks for the price per gallon of gasoline. The answer will be in dollars and cents (not in gallons!).

2. HOW—HOW do we solve the problem?

The problem asks for the price per gallon of gasoline. It is asking the price of each gallon. "Per" is a key word that tells us to divide. You have to take $16.50 and divide that amount into 15 parts. The operation is *divide*.

3. HELP—What HELP do we need?

You need a division setup such as:

$$15\overline{)16.50}$$

4. COMPUTE—divide.

Here is the computation:

$$
\begin{array}{r}
1.10 \\
15\overline{)16.50} \\
-15 \\
\hline
15 \\
-15 \\
\hline
0
\end{array}
$$

The answer is $1.10 per gallon.

5. CHECK—There are two CHECKS:

Part 1: CHECK to see if the answer is reasonable. Does it check out with your common sense?

If the answer came out to something like $100 per gallon, we would be very suspicious. Or if it were 10 cents per gallon, we would look at it strangely. When you divide $16.50 into 15 parts, you expect a dollar and some change.

Part 2: The second CHECK involves multiplying the answer 1.10 by 15, that is, by reversing the operation of division:

$$
\begin{array}{r}
1.10 \\
\times \quad 15 \\
\hline
5\ 50 \\
11\ 0 \\
\hline
16.50
\end{array}
$$

The answer $16.50 is the number we started with, so Part 2 works out OK.

✎ Exercises

For these questions use this next problem to get the answers.

Julie measured a plant in the spring. It was 32 cm high. When she measured it in the fall, it was 78 cm tall. How much did the plant grow?

Choose the correct answer to each question.

1. WHAT does the problem want us to find?

   a. the height of the plant in the fall
   b. the amount the plant grew
   c. the height in the spring

2. HOW: Which word is a key word of the problem?

   a. measured
   b. tall
   c. grow

3. HELP: What kind of computation setup will solve this problem?

   a. addition
   b. subtraction
   c. multiplication

4. COMPUTE: What is a reasonable estimate for the answer to the problem?

   a. about 50 cm
   b. about 100 cm
   c. about 120 cm

5. CHECK: What operation would you use to CHECK the computation?

   a. addition
   b. subtraction
   c. division

1.11 Using the Five Steps—II

Here is another example that shows how to use the five steps.

EXAMPLE

Sondra was overjoyed. Her uncle's will says that he left her $\frac{2}{5}$ of his property. The property consists of 250 acres of vacant land. How much land will Sondra receive from her uncle's estate?

Read the problem over again to make sure you understand it.

Here are the five steps.

1. WHAT?—WHAT does the problem want you to find?

The problem asks for the amount of land that Sondra will receive from her uncle's estate.

2. HOW?—HOW can you solve the problem? What operation will solve the problem?

The operation is multiplication, because we will have to find $\frac{2}{5}$ of the property.

The word "of" is often a clue to multiply, particularly in phrases such as "$\frac{2}{5}$ of ".

3. HELP—What HELP do you need?

First, a diagram might help, even if you don't use it in the computation.

 Each section is $\frac{1}{5}$ of the land

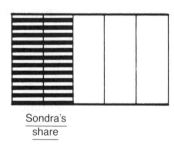

Sondra's
share

The diagram helps us visualize the size of the property that Sondra will receive.

Next, write an equation:

$$\frac{2}{5} \times 250 = ?$$

This equation is the computation setup.

4. COMPUTE—

$$\frac{2}{5} \times 250 \quad =$$

$$\underset{\$}{\underline{2 \times \overset{50}{\cancel{250}}}} \quad = 100$$

The answer is 100 acres. The area of the property is measured in acres, so the answer is also measured in acres.

5. CHECK—CHECK to see if the answer is reasonable.

Since we are taking a fraction ($\frac{2}{5}$) of the land, the answer has to be less than what we started with. Look at the diagram again.

Each section is $\frac{1}{5}$ of the total area. Each section is 50 acres. Two sections make up 100 acres. Does the answer make sense to you?

✏ Exercises

Use this problem to answer the questions.

Rita spent $5.25 for 25 pens. What is the cost of each pen?

Choose the correct answer for each question.

1. What does the problem want us to find?

    a. the amount of money that Rita spent
    b. the number of pens that Rita bought
    c. the cost of each pen

2. Which word in the problem is a key word?

    a. each
    b. spent
    c. pens

3. What operation will solve the problem?

    a. addition
    b. subtraction
    c. division

4. What is the answer to the problem?

    a. 15 cents
    b. 21 cents
    c. 25 cents

5. What operation would you use for a check of the computation?

    a. addition
    b. multiplication
    c. division

2. Whole Numbers

2.1 Whole Numbers: Getting Started

This chapter is about whole numbers. In this chapter you will apply the five-step strategy that you learned in Chapter 1 to solving problems with whole numbers.

The main skills you have to know for solving problems in this section are:

1. Adding whole numbers.

2. Subtracting whole numbers.

3. Multiplying whole numbers.

4. Dividing whole numbers.

If you are rusty with any of these skills, review them in this section before you go on.

Do you remember what parentheses mean in a problem? If an exercise has parentheses then the computation inside the parentheses must be done first.

Example: $(68 - 19) \times 72$

What to do: First subtract 19 from 68, and then multiply that answer by 72.

✎ Exercises

Which of these is a whole number? Draw a circle around each of the whole numbers.

1. $\frac{2}{5}$

2. 237

3. 0.235

4. 2,000,000

5. $\frac{5}{2}$

6. $6\frac{3}{4}$

7. 2

8. 1988

9. 13.75

10. 0

11. -23

12. 235.003

13. 141,111,111.2

14. 300

Explain how you would get the answer to these computations. Do not work out the answer.

15. $\dfrac{(64 + 87)}{2}$ _____

16. $9(872 + 451)$ _____

17. $(34 \times 5) + 18$ _____

18. (321 − 3) + 46 _____

19. 104 × (17 × 84) _____

20. (1045 ÷ 875) − 342 _____

21. Which of these is not a whole number?

    a. 1708

    b. 4.235

    c. 1

    d. 78

22. What operation does the box stand for?

        56 ☐ 23 = 1288

    a. addition

    b. subtraction

    c. multiplication

    d. division

23. What operation does the box stand for?

        405 ☐ 128 = 277

    a. addition

    b. subtraction

    c. multiplication

    d. division

24. What operation does the box stand for?

        648 ☐ 9 = 72

    a. addition

    b. subtraction

    c. multiplication

    d. division

Solve each equation.

25. 235 + a = 455 _____

26. 782 − b = 201 _____

27. 23 × d = 1035 _____

28. 432 − f = 24 _____

2.2 Whole Numbers: Examples

The five-step strategy for solving problems was introduced in Chapter 1. In this section you will see two examples that use the five steps to get the solution. If you have to, check back in Chapter 1 to see a full explanation of each step. In particular, refer to Sections 1.10 and 1.11, where you will find detailed explanations of the five steps applied to examples. Here is a quick review.

The method consists of five steps:

WHAT　　HOW　　HELP　　COMPUTE　　CHECK

To help you remember these steps, we introduced a memory gem sentence in Chapter 1:

WHAT! HOWARD HELPS COMPUTE CHECKS?

Each word in the sentence contains the name of one of the steps.

Now here are two examples showing how to find the answers. Follow the way we use the five steps to get the answer.

- -

EXAMPLE 1　　Phil scored 116 points in the first half of the season and 178 points in the second half of the season. How many points did he score altogether?

SOLUTION

Step 1. WHAT?　　Find the points Phil scored in the season.

Step 2. HOW?　　The word "altogether" gives us clue that we have to add (see Section 1.9—Key Words). Add the points of the first half of the season to the points of the second half of the season.

Step 3. HELP:　　Here is the setup for computation:

$$\begin{array}{r} 116 \text{ points} \\ +\,178 \text{ points} \\ \hline ? \end{array}$$

Step 4. COMPUTE: Add:

$$\begin{array}{r} 116 \text{ points} \\ +\,178 \text{ points} \\ \hline 294 \text{ points} \end{array}$$

The answer is 294 points.

Step 5. CHECK: You can check this answer by subtracting:

$$\begin{array}{r} 294 \\ -\,178 \\ \hline 116 \end{array}$$

- -

EXAMPLE 2   The temperature at 11 a.m. was 35 degrees 35°F (35 degrees Fahrenheit). It was 28° at 4 p.m. How many degrees did the temperature drop from 11 a.m. to 4 p.m.?

SOLUTION

Step 1: WHAT?   Find the temperature difference between 11 a.m. and 4 p.m.

Step 2. HOW?   The clue in this problem is "drop." This word suggests subtract.

Step 3. HELP: This is the setup for subtraction:

$$\begin{array}{r} 35° \\ -28° \\ \hline ? \end{array}$$

Step 4. COMPUTE: Subtract:

$$\begin{array}{r} 35° \\ -28° \\ \hline 7° \end{array}$$

The answer is 7°.

Step 5. CHECK: Add to check this answer:

$$\begin{array}{r} 28 \\ +\ 7 \\ \hline 35 \end{array}$$

Study these problems. In the next section you will get a chance to practice solving problems.

✎ Exercises

1. Which operation does the key word "altogether" suggest in Example 1?

   a. addition
   b. subtraction
   c. multiplication
   d. division

2. Which operation does the word "drop" suggest in Example 2?

   a. addition
   b. subtraction
   c. multiplication
   d. division

3. How was Example 1 above checked?

   a. by adding
   b. by subtracting
   c. by multiplying
   d. by dividing

4. How was Example 2 above checked?

    a. by adding
    b. by subtracting
    c. by multiplying
    d. by dividing

5. In which step do we get the answer?

    a. Step 2
    b. Step 3
    c. Step 4
    d. Step 5

6. Which question is answered by the first step?

    a. What does the problem want us to find?
    b. What way can we get help to solve the problem?
    c. How do we check the answer?

7. Match the words of the first column with the words of the second column. Draw a line connecting the word on the left to the word or words on the right.

WHAT?	REASONABLE?
COMPUTE	SET-UP or EQUATION
CHECK	CALCULATION
HOW?	QUESTION
HELP	KEY WORD

2.3 How Many Computation Steps?

You have learned a strategy to help you whenever you have to do a word problem. But there is one more thing to consider. When you get to the HOW step of the strategy, you must be prepared to ask the question:

How many computation steps will it take to solve the problem?

Many problems require only one computation step. Sometimes you will only have to add, or you will only have to multiply to solve the problem. But some problems require more than one step. For example, you may have to multiply and then add to get the answer.

You need to recognize if a problem requires only one computation step, or more than one. This is part of the overall strategy. Remember, you figure out the number of computation steps in the HOW step.

Here are two examples. The first one requires one computation step, the second one two computation steps.

EXAMPLE 1: ONE COMPUTATION STEP

Jane worked 187 hours last month, but she worked only 92 hours this month. How many more hours did she work last month?

SOLUTION

Step 1. WHAT?   Find out how many more hours Jane worked this month.

Step 2. HOW?   The clue is "more." The word suggests subtract. But how many computation steps? It takes only one computation step to get the answer.

Step 3. HELP   Here is the setup:

$$\begin{array}{r} 187 \text{ hours} \\ - \ 92 \text{ hours} \\ \hline ? \end{array}$$

Step 4. COMPUTE   Subtract:

$$\begin{array}{r} 187 \text{ hours} \\ - \ 92 \text{ hours} \\ \hline 95 \text{ hours} \end{array}$$

The answer is 95 hours.

Step 5. CHECK   Add to check the answer.

$$\begin{array}{r} 95 \\ + 92 \\ \hline 187 \end{array}$$

EXAMPLE 2: TWO COMPUTATION STEPS

Vanessa typed 22 letters each day for 8 days. Her friend Kevin typed 203 letters during the same time. How many fewer letters did Vanessa type?

SOLUTION

Step 1. WHAT?   Find out how many fewer letters Vanessa typed.

Step 2. HOW?   The clue is "fewer." The word suggests subtract. But how many computation steps? This problem requires two steps to get the answer.

Step 3. HELP   Here is the setup for the two steps:

Step 1

(How many letters Vanessa typed)

$$22 \\ \times 8$$

Step 2

(How many fewer she typed than Kevin)
Take answer of Step 1 and
subtract from 203

Step 4. COMPUTE   Multiply and subtract:

Step 1

22 letters
$\times 8$
176 letters

Step 2

203 letters
-176 letters
27 letters

The answer is 27 letters.

Step 5. CHECK Two parts—divide and add:

$$\begin{array}{r} 22 \\ 8)\overline{176} \end{array} \qquad \begin{array}{r} 27 \\ +\ 176 \\ \hline 203 \end{array}$$

Study these examples to make sure you understand why it is important to recognize the number of computation steps.

✏ *Exercises*

For each of these problems find the number of computation steps required to find the answer. <u>Do not work out the answers.</u> Your answers can be 1, 2, or 3 steps.

1. Sue charges $6 for each car she washes. If she washes a station wagon, she charges $8. How much did she earn when she washed 4 cars and 5 station wagons?

Number of steps to find the answer: _____

2. Lisa brought 68 balloons to the party this year. This was 14 more than last year. How many did she bring last year?

Number of steps to find the answer: _____

3. Phil filed 125 cards each day for the last 19 days. His goal is to reach 10,000. How many more does he have to file?

Number of steps to find the answer: _____

4. Maddy drove 245 miles the first day, and 120 miles for each of the next 5 days. How many miles did she drive in the 6 days?

Number of steps to find the answer: _____

5. Cathy figures she ate 2300 calories on Wednesday, 2400 on Thursday and 1700 on Friday. How many calories did she eat during these 3 days?

Number of steps to find the answer: _____

6. Jody bought 36 baseball cards on Tuesday and added them to her collection of 240 cards. She gave the cards to 12 students. Each student received the same number. How many did each student receive?

Number of steps to find the answer: _____

2.4 Whole Numbers: Practice

This section provides practice in solving word problems with whole numbers—the topic of this chapter. Solve the problems using the five-step method. Be careful of problems that require more than one computation step. In the HOW of the strategy, figure out how many computation steps the problem requires.

1. For her party, Sharon bought 23 packs of soda. Each pack had 6 sodas. How many sodas did she buy?

Your answer: _____

2. At the beginning of the semester, there were 27 students in Manny's class. By Thanksgiving, there were 19 students. How many fewer students were in Manny's class by Thanksgiving?

Your answer: _____

3. There are three apartment buidlings on Seventh Street. The first building has 23 apartments, the second 31 apartments and the third building 18 apartments. How many apartments are there altogether?

Your answer: _____

4. There are 40 candies in a box. If Aliza eats 4 each day, how many days will it take her to eat all the candies?

Your answer: _____

5. How many minutes are there in three hours? (An hour has 60 minutes.)

Your answer: _____

6. Jack's car gets 28 miles per gallon. Juanita's car gets 19 miles per gallon. If each used 11 gallons, how many miles more did Jack drive than did Juanita?

Your answer: _____

7. Seth bought eight cans of tuna fish. Each can weighs 14 ounces. How many ounces of tuna fish did Seth buy? (Be careful—the word *each* is not a key word in this problem!)

Your answer: _____

8. Joanna's weekly salary is $325. The deductions are $72. What is her take home pay?

Your answer: $_____

9. Tony counted 456 eggs. How many dozen did he count? (There are 12 in a dozen.)

Your answer: _____

10. Sal bought shares in two different companies. He bought 25 shares of ABC Company at the price of $12 per share. He bought 75 shares of XYZ Company at the price of $14 per share. How much did he pay for all the shares?

Your answer: $_____

2.5 Whole Numbers: Checkout

The purpose of this section is to give you a chance to check out your skills in solving word problems with whole numbers. Don't forget to use the 5-step strategy.

After each problem, you will find four answers. Select the correct answer for the problem by circling the letter next to the answer.

Sometimes, one of the choices will be NOT GIVEN. This choice means that the correct answer is not given among the other three choices. If you can't find the correct answer, then you should circle the letter next to NOT GIVEN.

1. Fred drove 177 miles in three days. How many miles did he drive each day if he drove the same distance each day?

   a. 55
   b. 59
   c. 354
   d. 531

2. Tina's gross salary each week is $505. Henry's gross salary is $450. How much money do they earn together each week?

   a. $55
   b. $885
   c. $955
   d. NOT GIVEN

3. Kristina does 15 situps each day. How many does she do in 2 weeks?

   a. 105
   b. 150
   c. 200
   d. NOT GIVEN

4. Judd owns 510 acres of land. He wants to divide the land into 15 equal parts. How large will each part be?

   a. 34 acres
   b. 36 acres
   c. 42 acres
   d. 75 acres

5. There were 3,568 people at the first football game. At the second game there were 6,859 people. How many more people were there at the second game?

   a. 292
   b. 2291
   c. 3291
   d. NOT GIVEN

6. In August, Seth jogged 87 miles. In September, he jogged 92 miles. How many miles did he run in these two months?

   a. 154
   b. 169
   c. 179
   d. 259

7. Luther drives 17 miles to work each day. After work he takes a short cut home which is 14 miles. How many miles does he drive to and from work each week (5 days)?

   a. 105
   b. 155
   c. 217
   d. NOT GIVEN

8. The Classical Music Shop has 812 tapes in stock. The Rock Music Shop has 12 times as many tapes as the Classical Music Shop. How many tapes are there in the Rock Music Shop?

   a. 800
   b. 974
   c. 9744
   d. NOT GIVEN

9. There were 174 people waiting on line this morning for tickets to the concert. Yesterday morning there were six times as many people. How many people were on line yesterday?

   a. 104
   b. 1040
   c. 14400
   d. NOT GIVEN

10. The population of Zimbabwe is 7,740,000 and the population of the United Arab Emirates is 1,210,000. How many more people are there in Zimbabwe?

    a. 6,530,000
    b. 6,520,000
    c. 8,950,000
    d. NOT GIVEN

3. *Fractions*

3.1 Fractions: Getting Started

This chapter is about solving word problems that contain fractions. The main skills that you need for this chapter are:

- Ability to work with equivalent fractions such as $\frac{2}{4} = \frac{1}{2}$. You need this skill to reduce fractions to their lowest terms.

- Finding a common denominator.

- Adding, subtracting, multiplying and dividing fractions.

If you have forgotten some of your fraction skills, review those skills before you go any further into this chapter. Here is a short reminder of the four operations with fractions.

Adding Fractions

Example $\frac{2}{3} + \frac{1}{4} = ?$

Here's how to add: Find the lowest common denominator. It is 12.

Change both fractions to 12ths:

$$\frac{2}{3} = \frac{8}{12} \text{ and } \frac{1}{4} = \frac{3}{12}.$$

Then add the numerators:

$$\frac{8}{12} + \frac{3}{12} = \frac{11}{12}$$

Subtracting Fractions

Example $4\frac{7}{8} - 2\frac{1}{4} = ?$

Here's how to subtract: Since the fractions are mixed fractions, place them under each other. Then find the lowest common denominator for the two fractions. Subtract the fractions and subtract the whole numbers:

$$
\begin{array}{rcr}
4\dfrac{7}{8} & = & 4\dfrac{7}{8} \\[2ex]
-2\dfrac{1}{4} & = & -2\dfrac{4}{8} \\[1ex]
\hline
& & 2\dfrac{3}{8}
\end{array}
$$

Multiplying fractions

Example $\dfrac{3}{7} \times \dfrac{4}{5} = ?$

Here's how to multiply: multiply the numerators and multiply the denominators:

$$\frac{3}{7} \times \frac{4}{5} = \frac{12}{35}$$

Dividing fractions

Example $\dfrac{2}{3} \div \dfrac{4}{7} = ?$

Here's how to divide: find the reciprocal of the second fraction. (This means to turn it upside-down— $\dfrac{4}{7}$ becomes $\dfrac{7}{4}$.) Then multiply:

$$\frac{2}{3} \times \frac{7}{4} = \frac{14}{12} \quad \text{or} \quad \frac{7}{6} \text{ or } 1\frac{1}{6}$$

We hope the following fact gives you some relief: division comes up much less frequently in word problems than the other three operations.

✎ Exercises _____

Circle the fractions.

1. $\dfrac{2}{3}$

2. 70,054

3. $\dfrac{7}{5}$

4. $\dfrac{1}{100}$

5. 6.77

6. $\dfrac{1}{32}$

7. 0

8. 2.5

Reduce these fractions to lowest terms.

9. $\dfrac{6}{12}$ = _____

10. $\dfrac{9}{27}$ = _____

11. $\dfrac{10}{200}$ = _____

12. $\dfrac{13}{65}$ = _____

13. $\dfrac{56}{7}$ = _____

14. $\dfrac{119}{17}$ = _____

Select the correct answer for each of the following.

15. What is the first step when multiplying these two fractions:

$$\frac{4}{5} \times \frac{6}{7} = ?$$

a. add 4 and 6
b. add 5 and 7
c. multiply 4 and 6
c. multiply 4 and 7

16. To divide these two fractions

$$\frac{2}{3} \div \frac{4}{5} = ?$$

one way to get the answer is to:

a. multiply $\frac{2}{3}$ by $\frac{4}{5}$

b. multiply $\frac{2}{3}$ by $\frac{5}{4}$

c. add $\frac{2}{3}$ and $\frac{5}{4}$

d. subtract $\frac{2}{3}$ from $\frac{5}{4}$

17. When you add these two fractions

$$\frac{2}{3} + \frac{5}{12} = ?$$

what is the first step?

a. Find common numerator
b. Find common denominator
c. Add the numerators
d. Add the denominators

18. When you subtract these fractions:

$$\frac{3}{7} - \frac{4}{5} = ?$$

common denominator is:

a. 12
b. 15
c. 28
d. 35

19. Dividing $\frac{3}{4}$ by $\frac{2}{3}$ is the same as:

    a. multiplying $\frac{3}{4}$ by $\frac{2}{3}$

    b. dividing $\frac{2}{3}$ by $\frac{3}{4}$

    c. multiplying $\frac{3}{4}$ by $\frac{3}{2}$

    d. subtracting $\frac{2}{3}$ from $\frac{3}{4}$

20. Add: $\frac{4}{5} + \frac{5}{4} = ?$ _____

21. Subtract: $\frac{7}{8} - \frac{1}{6} = ?$ _____

22. Multiply: $2\frac{1}{2} \times \frac{3}{4} = ?$ _____

23. Divide $\frac{1}{8} \div \frac{3}{4} = ?$ _____

3.2 Fractions: Examples

In this section, there are two examples of word problems with fractions. Both are worked out using the 5-step strategy.

Here is a quick review of the 5-step strategy for solving word problems with fractions. For a full explanation of the strategy, see Chapter 1. These are the five steps of the strategy:

WHAT   HOW   HELP   COMPUTE   CHECK

To help you remember these five steps, we introduced a memory gem:

WHAT! HOWARD HELPS COMPUTE CHECKS?

Each word in this sentence contains the name of one of the steps. Remember this sentence, and use the five steps when you solve word problems.

Study the way we use the steps in these two examples.

- -

EXAMPLE 1 Jean bought 24 tickets for the concert. Her friend Steve bought $\frac{1}{2}$ as many as she did. How many tickets did Steve buy?

SOLUTION

Step 1. WHAT? Find how many tickets Steve bought

Step 2. HOW? The words "$\frac{1}{2}$ as many as" suggest that we multiply by $\frac{1}{2}$

Step 3. HELP Here is the setup that will help get the answer:

$$\frac{1}{2} \times 24 = ?$$

Step 4. COMPUTE Multiply.

$$\frac{1}{2} \times 24 = ?$$

Change 24 to $\frac{24}{1}$ and multiply:

$$\frac{1}{2} \times \frac{24}{1} = \frac{24}{2} = 12 \text{ tickets}$$

The answer is 12 tickets.

Step 5. CHECK   To check, first ask: is this answer reasonable? Then, check the calculation by multiplying by 2:

$$12 \times 2 = 24$$

- -

EXAMPLE 2 Fran and Tom are baking the crust for a pie. The recipe calls for $\frac{2}{3}$ cup of sugar and $\frac{3}{4}$ cup of flour. How much sugar and flour is that altogether?

SOLUTION

Step 1. WHAT? Find how much sugar and flour altogether

Step 2. HOW? The word "altogether" suggests that we add. (Multiplying wouldn't make sense here.)

Step 3. HELP The setup looks like this:

$$\frac{2}{3} + \frac{3}{4} = ?$$

Step 4. COMPUTE To add, we need to find a common denominator.

The least common denominator for 3 and 4 is 12; so:

$$\frac{2}{3} = \frac{8}{12} \quad \text{and} \quad \frac{3}{4} = \frac{9}{12}.$$

$$\frac{8}{12} + \frac{9}{12} = \frac{17}{12} = 1\frac{5}{12} \text{ cups}$$

Step 5. CHECK To check your answer, subtract:

$$\frac{17}{12} - \frac{9}{12} = \frac{8}{12}$$

Go over these examples. In the next section, you will get a chance to practice solving fraction word problems.

✎ Exercises

1. What is the numerator of $\frac{7}{9}$?

2. What is the denominator of $\frac{13}{100}$?

Choose the answer to each of the following.

3. Which fractions are less than one?

   a. $\frac{1}{7}$ **d.** $\frac{4}{3}$

   b. $\frac{16}{16}$ **e.** $\frac{2}{3}$

   c. $\frac{23}{24}$ **f.** $\frac{24}{23}$

4. Which of these is equivalent to $\frac{2}{3}$?

   a. $\frac{4}{6}$ **c.** $\frac{4}{3}$

   b. $\frac{2}{6}$ **d.** $\frac{4}{12}$

5. Which is another way to write 35?

 a. $\dfrac{35}{2}$

 b. $\dfrac{1}{35}$

 c. $\dfrac{35}{1}$

 d. $\dfrac{70}{1}$

6. In which step of the five-step strategy do we ask: does this answer make sense?

 a. WHAT?
 b. HOW?
 c. HELP
 d. CHECK

7. In which step do we use the computational setup?

 a. WHAT?
 b. HOW?
 c. COMPUTE
 d. CHECK

8. In which step do we ask: what does the problem want us to find?

 a. WHAT?
 b. HOW?
 c. HELP
 d. CHECK

9. Which is the third step of the strategy?

 a. WHAT?
 b. CHECK
 c. COMPUTE
 d. HELP

10. In which step would you draw a diagram?

 a. HELP
 b. COMPUTE
 c. WHAT?
 d. HOW?

3.3 How Many Computation Steps?

When you use the 5-step strategy, there is an important thing to remember in the HOW step. You must ask the question:

How many computation steps will it take to solve the problem?

Many problems require only one computation step. Sometimes you will only have to subtract, or you will only have to multiply to solve the problem. But some problems take more than one step. For example, you may have to multiply and then add to get the answer.

You need to recognize if a problem requires one computation step, or if the problem requires more than one computation step. You do that in the HOW step. It is part of the strategy.

Here is an example with two computation steps.

EXAMPLE: TWO COMPUTATION STEPS

Before this week, John worked $13\frac{1}{2}$ days on his new job. He will work $\frac{1}{2}$ of the regular $4\frac{1}{2}$ days this week. How many days will he have worked after this week?

SOLUTION

Step 1. WHAT?   Find out how many days John has worked.

Step 2. HOW?   How many computation steps? Study the problem and you will see that it requires two steps to get the answer. One clue is "of." The word suggests multiplication as one of the steps.

Step 3. HELP   Here is the setup for the two steps:

Step 1	Step 2
$\frac{1}{2} \times 4\frac{1}{2} = ?$	Take answer of Step 1 and add it to $13\frac{1}{2}$

Step 4. COMPUTE   Multiply and add:

$$
\begin{array}{cc}
\underline{\text{Step 1}} & \underline{\text{Step 2}} \\
\frac{1}{2} \times 4\frac{1}{2} = ? & 13\frac{1}{2} \text{ days} \\
\frac{1}{2} \times \frac{9}{2} = \frac{9}{4} = 2\frac{1}{4} & +2\frac{1}{4} \text{ days} \\
 & \overline{15\frac{3}{4} \text{ days}}
\end{array}
$$

The answer is $15\frac{3}{4}$ days.

Step 5. CHECK Two parts—divide and subtract:

Check of Step 1

$$2\frac{1}{4} \div 4\frac{1}{2} = ?$$

$$\frac{9}{4} \div \frac{9}{2} - \frac{9}{4} \times \frac{2}{9} - \frac{1}{2}$$

Check of Step 2

$$15\frac{3}{4}$$
$$-\ 2\frac{1}{4}$$
$$\overline{13\frac{1}{2}}$$

Study this example to make sure you understand why it is important to recognize the number of computation steps.

✎ *Exercises*

For each of these problems find the number of computation steps required to find the answer. *Do not work out the answers.*

1. Henry calculated that $\frac{2}{3}$ of his summer vacation was over. If the entire vacation is 57 days, how many days are over?

   Number of computation steps: _____

2. Dina's friend owes her $200. Dina will get $\frac{1}{4}$ of the money at the end of the week. At that time, she will give $\frac{5}{8}$ of what she gets to her brother. How much will her brother get?

   Number of computation steps: _____

3. Dan started with a ribbon $4\frac{5}{8}$ inches long. He first cut a piece $\frac{7}{8}$ inch off. Then Then he cut $\frac{1}{4}$ of the remainder to decorate his room. How much did he take?

   Number of computation steps: _____

4. Ramon mixed $\frac{2}{3}$ pound of chocolate with $\frac{1}{2}$ pound of sugar. How much did these two ingredients weigh?

   Number of computation steps: _____

5. Angela had to drive 480 miles. She drove $\frac{1}{2}$ the distance the first day, $\frac{1}{3}$ the distance the second day and $\frac{1}{8}$ the distance the third day. How far did she get after 3 days?

   Number of computation steps: _____

3.4 Fractions: Practice

There are 11 problems in this section for you to practice. Use the 5-step strategy. Make sure you figure out the number of computation steps during the HOW step.

1. Joan lives 3 miles from school. One morning she jogged $\frac{5}{8}$ mile to school and walked $\frac{3}{4}$ mile. How much further did she have to go?

Your answer: _____ miles

2. Terry drove 19 miles in $\frac{1}{3}$ hour. What was her speed in miles per hour?

Your answer: _____

3. Sonja can walk $4\frac{1}{2}$ miles an hour. How many miles can she walk in $2\frac{1}{2}$ hours?

Your answer: _____

4. Bella usually spends 38 hours a week in school. Last week she was sick and missed $3\frac{1}{2}$ hours on Wednesday and $2\frac{3}{4}$ hours on Thursday. How many hours did she spend in school?

Your answer: _____

5. The price of one share of ABC Co. stock was listed as $18\frac{3}{4}$ at the beginning of the week. At the end of the week the price was listed as $22\frac{1}{8}$. How much did the price gain during the week?

Your answer: $ _____

6. Each day Stella buys $2\frac{1}{2}$ pounds of cheese. How much does she buy in 7 days?

Your answer: _____ pounds

7. Carol decided she was going to spend 10 hours on housework this weekend. She spent $3\frac{3}{4}$ hours painting her apartment and $2\frac{2}{3}$ hours cleaning up. How many more hours did she have left for housework?

Your answer: _____

8. There are 576 students in school. If $\frac{3}{4}$ of them do not exercise, then how many students exercise?

Your answer: _____

9. There were 72 people on the trip. Three fourths of them voted not to go to the opera. How many voted against going to the opera?

Your answer: _____

10. Larry needs 18 inches of ribbon. He has $9\frac{5}{8}$ inches of blue ribbon and $5\frac{1}{4}$ inches of green ribbon. How much more ribbon does he need?

Your answer: _____ inches

11. Simon divides $6\frac{2}{3}$ feet of wire into 3 parts. How long is each part?

Your answer: _____ feet

3.5 Fractions: Checkout

The purpose of this section is to give you a chance to check out your skills in solving fraction word problems. Don't forget to use the 5-step approach.

For each example, you will find four answers. Select the correct answer for the problem by circling the letter next to the answer.

Sometimes, one of the answers will be NOT GIVEN. This choice means that the correct answer is not given among the other three choices. If you can't find the correct answer, then you should circle the letter next to NOT GIVEN.

1. Jack is baking a cake. The recipe says that it must bake for $2\frac{1}{2}$ hours. It has been baking for $\frac{3}{4}$ hour. How much longer must it bake?

   a. $2\frac{1}{4}$ hours

   b. $1\frac{3}{4}$ hours

   c. $1\frac{1}{2}$ hours

   d. $\frac{3}{4}$ hour

2. Carla works $7\frac{1}{2}$ hours a day. She works 5 days each week. How many hours does she work each week?

   a. 34

   b. $35\frac{1}{2}$

   c. $37\frac{1}{2}$

   d. 38

3. Malcolm earned $75 raking leaves. He worked $5\frac{1}{2}$ hours Friday, 6 hours on Saturday and $3\frac{1}{2}$ hours on Sunday. How much money did he make per hour?

   a. $4 per hour

   b. $4\frac{1}{2}$ per hour

   c. $5 per hour

   d. $6 per hour

4. Harvey rode 15 miles on his bike in $1\frac{1}{2}$ hours. What was his speed?

   a. 8 miles per hour
   b. 10 miles per hour
   c. 12 miles per hour
   d. NOT GIVEN

5. Wanda spent $4\frac{3}{4}$ hours watching TV on Monday, and $3\frac{2}{3}$ hours on Tuesday. How many hours did she watch TV on these two days?

   a. $6\frac{5}{12}$
   b. $7\frac{5}{12}$
   c. $8\frac{1}{2}$
   d. NOT GIVEN

6. Tanya swims $\frac{5}{6}$ mile every day. How many miles does she swim every week (7 days)?

   a. $3\frac{1}{2}$
   b. $4\frac{5}{6}$
   c. $5\frac{1}{2}$
   d. NOT GIVEN

7. There were 20 people at Ralph's after school. Four-fifths of the people recognized Aliza when she walked in. How many people recognized Aliza?

   a. 4
   b. 16
   c. 18
   d. NOT GIVEN

8. Ivan ate $\frac{1}{3}$ of the pizza before he played soccer. After he played, he ate another $\frac{2}{5}$ of the pie. What part of the pizza did he eat altogether?

    a. $\frac{3}{5}$

    b. $\frac{4}{15}$

    c. $\frac{11}{15}$

    d. NOT GIVEN

9. Pedro can write 2 pages in $\frac{2}{3}$ hour. At that rate, how many pages can he write in 1 hour?

    a. 1

    b. $2\frac{1}{2}$

    c. 4

    d. NOT GIVEN

10. The roast beef that Tom is cooking takes $4\frac{1}{4}$ hours. The roast has been cooking for $2\frac{3}{4}$ hours. How much longer should it cook?

    a. $2\frac{2}{3}$ hours

    b. $2\frac{1}{2}$ hours

    c. $1\frac{3}{4}$ hours

    d. NOT GIVEN

4. Decimals

4.1 Decimals: Getting Started

In this chapter, you will study how to solve problems with decimals, including problems with dollars and cents. The main skills that you need in this chapter are the following:

1) Relating decimals to fractions such as $0.3 = \dfrac{3}{10}$
2) Putting decimals in order.
3) Adding, subtracting, multiplying, and dividing decimals.

Review these skills if you have forgotten them. The exercises below will help refresh your memory on some of the skills.

✎ Exercises

Circle the decimal numbers.

1. 0.56

2. 6.24

3. 100.08

4. 2.5

5. $\dfrac{7}{8}$

6. 90,000

7. $7\dfrac{3}{4}$

8. 0.00095

Change these decimals to fractions.

9. 0.25

10. 0.13

11. 5.5

12. 13.567

Change these fractions to decimals.

13. $\dfrac{9}{10}$

14. $\dfrac{1}{2}$

15. $\dfrac{2}{5}$

16. $2\dfrac{3}{1000}$

17. $1\dfrac{3}{4}$

18. $56\dfrac{5}{8}$

19. Add: 2.345 + 100.67

20. Multiply: 7.6 × 8.034

21. Subtract: 2.004 − 1.871

22. Divide: 23.56 ÷ 1.4 (to 2 decimal places)

Circle the larger of the two decimals in each pair.

23. **a.** 0.235   **b.** 1.235

24. **a.** 2.025   **b.** 2.205

25. **a.** 10.761   **b.** 10.76

26. **a.** 100.099 **b.** 100.1

4.2 Decimals: Examples

In this section, you will find two examples using decimals. Both examples use the 5-step procedure and are fully worked out.

The 5-step procedure was introduced in Chapter 1. These five steps will make problem solving easier for you. Here is a quick review of the five steps:

Step 1. WHAT?
Step 2. HOW?
Step 3. HELP
Step 4. COMPUTE
Step 5. CHECK

In Chapter 1, we introduced this memory gem sentence to help you remember the five steps:

WHAT! HOWARD HELPS COMPUTE CHECKS?

Each word in the sentence contains the name of one of the steps.

Study how each step contributes to finding and checking the answer.

- -

EXAMPLE 1

Jodi spent $34.75 on blank computer disks. She also spent $45.25 on software. How much did she spend altogether?

SOLUTION

Step 1. WHAT?   Find how much Jodi spent.

Step 2. HOW?   The word "altogether" suggests that we add.

Step 3. HELP   Here is the setup that will help get the answer:

$$\begin{array}{r} \$34.75 \\ + \ 45.25 \\ \hline ? \end{array}$$

Step 4. COMPUTE Add:

$$\begin{array}{r} \$34.75 \\ + \ 45.25 \\ \hline \$80.00 \end{array}$$

The answer is $80.

Step 5. CHECK   To check this answer, subtract:

$$\begin{array}{r} \$80.00 \\ - \ 45.25 \\ \hline \$34.75 \end{array}$$

EXAMPLE 2

Clark went shopping at the supermarket. The bill came to $34.23. Clark paid for the groceries with a 50-dollar bill. How much change did he receive?

SOLUTION

Step 1. WHAT?   To find the amount of change

Step 2. HOW?   The words "How much change" suggest that we subtract.

Step 3. HELP   The setup looks like this:

$$\begin{array}{r} \$50.00 \\ -\ 34.23 \\ \hline ? \end{array}$$

Step 4. COMPUTE Subtract

$$\begin{array}{r} \$50.00 \\ -\ 34.23 \\ \hline \$15.77 \end{array}$$

The answer is $15.77.

Step 5. CHECK   To check your answer add:

$$\begin{array}{r} \$15.77 \\ +\ 34.23 \\ \hline \$50.00 \end{array}$$

Study these examples. In the next section, you will get a chance to practice solving word problems with decimals.

✎ *Exercises* _____

1. How many decimal places are there in the answer of 4.237×9.871?_____

2. Multiply: $2.358 \times 1000 =$ _____

3. Divide: $654.783 \div 100 =$ _____

4. Add: $2.548 + 84.6 + 0.0001 = ?$ _____

5. Subtract: $2.3819 - 1.1 = ?$ _____

Place the decimals in order from smallest to largest:

6. 2.149, 2.419, 2.2 _____

7. 7.091, 7.912, 7.090 _____

8. 0.012, 0.01, 0.011 _____

Choose the correct answers:

9. What operation do the words "How much change?" suggest?

    a. addition
    b. subtraction
    c. multiplication
    d. division

10. In which step of the 5-step procedure, do we ask: what do we have to find?

    a. WHAT?
    b. HOW?
    c. HELP
    d. CHECK

11. In which step do we make sure that the computation is correct?

    a. WHAT?
    b. HOW?
    c. HELP
    d. CHECK

12. In which step do we find the answer?

    a. WHAT?
    b. HOW?
    c. HELP
    d. COMPUTE

13. In which step do we figure out which operation to use?

    a. WHAT?
    b. HOW?
    c. HELP
    d. COMPUTE

14. In which step do you ask if the answer is reasonable?

    a. WHAT?
    b. HELP
    c. COMPUTE
    d. CHECK

4.3 How Many Computation Steps?

In the HOW step of the strategy you must be prepared to ask the question:

How many computation steps will it take to solve the problem?

Many problems require only one computation step—you may have to add, or you may have to multiply. But some problems require more than one computation step.

Here is an example that requires two computation steps.

EXAMPLE: TWO COMPUTATION STEPS

Sam brought 3 computer disks at $20.45 each and a video tape at $29.95. How much did Sam pay in all?

SOLUTION

Step 1. WHAT? Find how much Sam paid.

Step 2. HOW? The clue is "in all". These words suggest addition. But there is a multiplication step before addition. There are two steps to get the answer.

Step 3. HELP Here is the setup for the two steps:

Step 1	Step 2
$20.45	Add the answer of
× 3	Step 1 and $29.95

Step 4. COMPUTE Multiply and subtract:

Step 1	Step 2
$20.45	$61.35
× 3	+ 29.95
$61.35	$91.30

The answer is $91.30.

Step 5. CHECK Two parts—subtract and divide:

$$\begin{array}{r} \$91.30 \\ -\ 29.95 \\ \hline \$61.35 \end{array} \qquad \begin{array}{r} \$20.45 \\ 3\ )\overline{\$61.35} \end{array}$$

✎ Exercises

For each of these problems find the number of computation steps required to find the answer. Do not work out the answers.

1. Doris works as a waitress in a restaurant. The largest food bill yesterday was $58.23. The tax was $3.49. How much was the total bill?

   Number of computation steps:_____

2. Frank brought 3 hammers at $5.95 each and 4 boxes of nails at $1.26 each. How much did Frank pay for these items?

   Number of computation steps:_____

3. Suzanne drove 26.8 miles on the first day, and $\frac{1}{2}$ that distance the second day. How much did she drive altogether?

   Number of computation steps:_____

4. Dan's average in history was 78.4. Martha's average was 82.9. How much higher was Martha's average?

   Number of computation steps:_____

5. Stan mailed 13 packages yesterday. Six of the packages weighed 2.3 pounds each. The remaining 7 packages weighed 3.4 pounds each. How much did the packages weigh altogether?

   Number of computation steps:_____

4.4 Decimals: Practice

There are 11 practice problems in this section. Solve them using the 5-step strategy. Make sure to check the answers.

1. Tim bought three tapes. Each cost $5.75. What was the cost of the three tapes?

Your answer: $ —————

2. Olga bought a shirt for $18.95 and a sweater for $45. How much change did she get from $70?

Your answer: $ —————

3. Jack drove his car for 5.2 hours. He drove at an average speed of 47.5 miles per hour. How many miles did Jack drive?

Your answer: —————

4. Paula's checking account had been $347.72. She deposited a check for $239.41. What was her new balance?

Your answer: $ —————

5. Gasoline costs 80.5 cents per gallon. How much does 12 gallons cost?

Your answer: —————

6. Heidi earned $24,360 last year. How much did she make each month?

Your answer: $ —————

7. A telephone call from Smithville to Somerville costs $.22 for the first minute and $.18 for each additional minute. How much does a 15-minute call cost?

Your answer: $ —————

8. At the beginning of a trip, the mileage gauge read 23,457.3 miles. After the trip, the mileage gauge read 24,092.8 miles. How many miles long was the trip?

Your answer: —————

9. Jennifer bought 3 tickets to the movies at $4.25 each. Then she bought popcorn and a soda for $1.75. If she started with $20, how much money did she have left?

Your answer: $ —————

10. The overtime rate at ABC Company is 1.5 times the regular hourly wage. If Bela makes $4.70 per hour, then how much does she make at the overtime rate?

Your answer: $ —————

11. Stan paid $3.85 for seven cups of coffee for himself and his friends. How much did each cup of coffee cost?

Your answer: $ —————

4.5 *Decimals: Checkout*

The purpose of this section is to give you a chance to check out your skills in solving decimal word problems. Use the 5-step approach.

For each example, you will find four answers. Select the correct answer for the problem by circling the letter next to the answer.

Sometimes, one of the choices will be NOT GIVEN. This choice means that the correct answer is not given among the other three choices. If you can't find the correct answer, then you should circle the letter next to NOT GIVEN.

1. Juanita drives to work. She pays $5.65 in tolls each day. How much does it cost her each week (five days) for tolls?

   a. $19.75
   b. $20.75
   c. $27.25
   d. $28.25

2. Willie collected money in his office for a party. On Monday he collected $24.00. On Tuesday he collected $54.75 and on Wednesday he collected $23.50. How much did he collect in those three days?

   a. $102.25
   b. $78.75
   c. $78.25
   d. NOT GIVEN

3. Gasoline was selling at $1.20 per gallon. How many gallons can a person get for $15?

   a. 11.5
   b. 12.5
   c. 13
   d. 13.5

4. Lori went to a concert. Her ticket cost $9.75. Last year she paid $8.25. How much less was the concert last year?

   a. $11.50
   b. $8.00
   c. $1.75
   d. $1.50

5. Sam paid $.75 for orange juice, $2.50 for a sandwich and $1.25 for a bag of nuts. How much change did he get from $10.00?

   a. $6.75
   b. $6.25
   c. $4.50
   d. $5.50

6. Wally drove 7.2 miles at an average speed of 50 miles per hour. How many miles did Wally drive?

   a. 350
   b. 355
   c. 360
   d. NOT GIVEN

7. Rosie's weekly salary is $332.50. She works 35 hours a week. How much does she make per hour?

   a. $9.50
   b. $9.75
   c. $10.25
   d. NOT GIVEN

8. The balance in Keith's bank account was $704.29. He then deposited a check for $237.98. What is his new balance?

   a. $466.31
   b. $476.31
   c. $932.27
   d. $942.27

9. Sherry bought 2.5 pounds of ice cream at $5.40 per pound. How much change did she get from a 20 dollar bill?

   a. $5.50
   b. $6.50
   c. $13.50
   d. $14.50

10. A call from St. Louis to Kansas City costs $.37 for the first minute and $.23 for each additional minute. How much would a 13 minute call cost?

   a. $2.76
   b. $3.13
   c. $4.44
   d. $4.67

5. REVIEW TEST A

This test covers the topics of Chapters 2, 3 and 4. It consists of 20 questions on all three topics—whole numbers, fractions, and decimals.

There are four answers for each question. Select the correct answer to the problem by circling the letter next to the answer.

For some of the problems, one of the choices is NOT GIVEN. This choice means that the correct answer is not given among the other three choices. If you can't find the correct answer, then circle the letter next to NOT GIVEN.

1. Joan works at the local computer store. Her net weekly pay comes to $561.87. Her deductions amount to $241.73. How much is her gross pay each week?

   a. $320.14
   b. $703.60
   c. $803.60
   d. NOT GIVEN

2. Gordon worked 60 hours last week. If $\frac{3}{4}$ of the time he used the computer, how many hours did he use the computer?

   a. 15
   b. 30
   c. 45
   d. NOT GIVEN

3. There were 45,872 people at the stadium who wanted tickets to the football game. But only 8,561 of these people got tickets. How many people at the stadium did not get tickets?

   a. 36,311
   b. 37,311
   c. 53,433
   d. 54,433

4. Omar waited $\frac{1}{3}$ of an hour at the dentist's office. Then the dentist saw him for $\frac{1}{4}$ of an hour. How long was Omar's visit to the dentist's office?

   a. $\frac{1}{7}$ hour

   b. $\frac{1}{6}$ hour

   c. $\frac{2}{7}$ hour

   d. $\frac{7}{12}$ hour

5. Curtis spent $42.56 for records, $76.13 for clothing, and $18.34 for food. How much money did he have left from $200?

   a. $62.07
   b. $62.97
   c. $63.97
   d. $64.07

6. Willie works at the supermarket. He carried 17 boxes of canned tuna into the warehouse. Each box contains 48 cans of tuna. How many cans did he carry?

   a. 816
   b. 806
   c. 716
   d. NOT GIVEN

7. Pat's checkbook balance at the end of the month was $131.88. At the beginning of the month the balance was $67.23. Pat made a deposit of $400 during the month. How much money from the checking account did she spend during the month?

   a. $335.35
   b. $335.45
   c. $531.88
   d. $599.11

8. Jerry jogged $\frac{4}{7}$ of a mile in the morning and $1\frac{3}{7}$ miles in the afternoon. How far did he jog altogether?

   a. $2\frac{1}{7}$ miles
   b. 2 miles
   c. $1\frac{1}{7}$ miles
   d. NOT GIVEN

9. Dana got a new job. She will get paid $6.50 an hour. How much will she earn if she works 35 hours per week?

   a. $227
   b. $217.50
   c. $207.50
   d. NOT GIVEN

10. Fran places sodas into 6-packs. How many packs can she make if she starts with 924 sodas?

    a. 154
    b. 152
    c. 150
    d. NOT GIVEN

11. One-third of the senior class did not show up for the game. There are 276 students in the class. How many students did not show up?

a. 184
b. 138
c. 92
d. 91

12. George invests in the stock market. He owns 12 shares of TRY Computer Co. When he bought the shares 6 months ago, the price of each share was $14.25. The price now is $22.50. How much profit does he have from this investment?

a. $99.50
b. $99
c. $98.50
d. $98

13. Jill made a mistake. She used too much flour for the recipe. She used $2\frac{1}{4}$ cups of flour. The recipe calls for $1\frac{3}{4}$ cups. How much extra flour did she use?

a. $\frac{1}{4}$ cup
b. $\frac{1}{2}$ cup
c. 3 cups
d. NOT GIVEN

14. Louisa likes to shop. She started the day with $175. First, she bought three books at $6.85 each. Then she bought two videos at $22.75 each and finally she bought a coat for $85. How much money did she have left?

a. $141.05
b. $150.95
c. $151.15
d. NOT GIVEN

15. The population of California is approximately 25,000,000. The population of neighboring Nevada is approximately 891,000. About how many more people live in California than in Nevada?

a. 25,109,000
b. 24,219,000
c. 24,119,000
d. 24,109,000

16. Bjorn says a jar of coffee usually contains 72 tablespoons of coffee. Each morning he takes about $1\frac{1}{2}$ tablespoons out for his coffee. About how many mornings can Bjorn make coffee before the jar is empty?

    a. 108
    b. 48
    c. 24
    d. NOT GIVEN

17. The capacity of an auditorium is 487 seats. It was filled all three nights for the play. How many people saw the play?

    a. 1461
    b. 1361
    c. 1261
    d. NOT GIVEN

18. Five-sixths of the members of a condominium voted to increase the monthly charge to pay for a new pool table. There are 252 residences. How many people voted for the increase?

    a. 42
    b. 84
    c. 168
    d. 210

19. All 23 of the members of a club are going to split the profits evenly. There was $782 profit at the end of the season. How much did each member get?

    a. $23
    b. $28
    c. $33
    d. NOT GIVEN

20. Keith works at a coffee shop. He earns $65.40 per day. How much does he earn after six days?

    a. $372.40
    b. $372.90
    c. $392.40
    d. $392.90

6. *Percents*

6.1 *Percents: Getting Started*

In this chapter you solve problems with percents. As you know, to work with a percent, you first have to change it to either a decimal or a fraction. So these are the main skills that you need to know in order to solve problems with percents:

- How to change percents to decimals and fractions.
- How to perform operations with decimals and fractions.

Spend a little time reviewing these two main skills before you begin this chapter. Here is a quick look at them.

- -

Changing percents to fractions

Percent means "based on 100." So a percent translates directly into a fraction with 100 as a denominator. This fraction can often be reduced:

$$62\% = \frac{62}{100} = \frac{31}{50}$$

$$80\% = \frac{80}{100} = \frac{4}{5}$$

- -

Changing percents to decimals

Write the percent as a fraction first. (Don't reduce it.)

$$43\% = \frac{43}{100}$$

Then change the fraction to a decimal:

$$\frac{43}{100} = 0.43$$

The exercises below should help get you started.

✎ Exercises

Change each percent to a fraction.

1. 25% _____

2. 20% _____

3. 75% _____

4. 125% _____

Change each percent to a decimal.

5. 42% _____

6. 137% _____

7. 4% _____

8. 97% _____

9. 657% _____

10. 9% _____

Change each number to a percent.

11. $\dfrac{35}{100}$ _____

12. 0.80 _____

13. 4 _____

14. 1.5 _____

15. $\dfrac{3}{4}$ $\left(= \dfrac{?}{100}\right) =$ _____

16. $\dfrac{1}{2}$ _____

17. $\dfrac{72}{100}$ _____

18. $\dfrac{2}{5}$ _____

Compute:

19. 50% of 40 _____

20. 25% of 800 _____

6.2 *Percents: Examples*

Word problems with percents give many students trouble. It will help you to know that there are only three basic types. In this section of the chapter, we will give you examples of each type and show you how to solve each one.

You will still use the basic 5-step strategy you learned in Chapter 1:

> **Step 1.** WHAT?
> **Step 2.** HOW?
> **Step 3.** HELP
> **Step 4.** COMPUTE
> **Step 5.** CHECK

In Chapter 1 we introduced a memory gem sentence to help you remember the five steps:

WHAT! HOWARD HELPS COMPUTE CHECKS?

Each word in the sentence contains the name of one of the steps. Remember this sentence, and use the five steps to solve word problems. Study each step of the word problems in this chapter to see how the step contributes to finding the answer. Don't forget the important last step— checking the answer.

- -

Here are the three types of percent problems. The first example of each type explains the basic math. The second example gives you an actual word problem of the same type and shows you how to solve it.

TYPE 1: Finding a Percent of Another Number

EXAMPLE 1: How much is 40% of 300?

This is an example of the math in the first type of percent problem.

In this type, you are given the percent. You have to find that percent of another number which also appears in the problem.

SOLUTION: The word "of" means <u>multiply</u>. Here are two ways to do the problem:

1) Change 40% to a decimal and multiply:

$$\begin{array}{r} 300 \\ \times\ 0.40 \\ \hline 120 \quad \text{(answer)} \end{array}$$

2) Or, change 40% to a fraction and multiply:

$$40\% = \frac{40}{100} = \frac{4}{10} = \frac{2}{5}$$

$$\frac{2}{5} \times \frac{300}{1} = \frac{600}{5} = 120$$

Your answer is 120.

EXAMPLE 2: Last year Juan paid $450 for a new TV set. This year the same set sells for 8% less. How much did the price decrease from last year?

This is an example of a Type 1 word problem. You solve it the same way that you solve Example 1.

SOLUTION:

Step 1.	WHAT?	To find how much the price decreased.
Step 2.	HOW?	The words "8% less" mean "8% of $450 less." Change 8% to a decimal and multiply. (Be careful—"less" and "decrease" are not key words here.)
Step 3.	HELP	This is the computation setup:

$$\begin{array}{r} 450 \\ \times\ 0.08 \\ \hline ? \end{array}$$

| Step 4. | COMPUTE |

$$\begin{array}{r} 450 \\ \times\ 0.08 \\ \hline 36.00 \end{array}$$

The answer is $36.

| Step 5. | CHECK | To check this answer, divide 36 by 450 to get a percent: |

$$\frac{36}{450} = \frac{2}{25} = 0.08$$
$$= 8\%$$

TYPE 2: Finding What Percent One Number Is of Another

EXAMPLE 3: What percent is 14 of 56?

This is an example of the math in the second type of percent problem.

In this type, you want to find the percent when you are given two numbers.

SOLUTION: Form the fraction $\frac{14}{56}$ and change it to a percent:

$$\frac{14}{56} = 25\%$$

Your answer is 25%

EXAMPLE 4: Jill drove to see her cousin who lives 320 miles away. After she had driven 160 miles, what percent of the total trip had she completed?
This is an example of a Type 2 word problem. You solve it the same way you solve Example 3.

SOLUTION:

| Step 1. | WHAT? | To find the percent of the trip completed. |
| Step 2. | HOW? | What percent is 160 of 320? This question suggests Type 2. Type 2 says, form a fraction. |

Step 3. HELP   The setup looks like this:

$$\frac{160}{320}$$

Step 4. COMPUTE Reduce this fraction and then change the fraction to percent:

$$\frac{160}{320} = \frac{1}{2}$$
$$= 50\%$$

The answer is 50%.

Step 5. CHECK   To check your answer change 50% to a decimal or fraction and multiply:

Decimal: $320 \times .50 = 160$

Fraction: $320 \times \dfrac{1}{2} = 160$

TYPE 3: Finding a Number When a Percent of It Is Known

EXAMPLE 5: If 20% of a number is 13, what is the number?

This is an example of the math in the third type of percent problem.

In this type, you know the percent and the answer after you have taken the percent. You have to find the original number.

SOLUTION: We translate this question into an equation. The word "of" means "multiply," and the word "is" stands for "equals." Change 20% to a decimal and write an equation. Use *n* for the missing number:

$$.20 \times n = 13$$

You can solve this equation by dividing both sides by .20:

$$n = \frac{13}{.20} = \frac{130}{2}$$
$$= 65$$

Your answer is 65.

EXAMPLE 6: Jim received $180 in interest last year on an investment. If the investment paid 9%, how much was the original investment?

This is an example of a Type 3 word problem.

SOLUTION

Step 1. WHAT?   To find the amount of the original investment

Step 2. HOW?   Read the problem as "If 9% of an amount is $180, then what is the amount?" This is Type 3. Write an equation.

Step 3. HELP   Here is the equation:

$$.09 \times n = 180$$

Step 4. COMPUTE Divide both sides of the equation by .09

$$n = \frac{180}{.09} = \frac{18000}{9}$$
$$= 2000$$

The answer is $2000.

Step 5. CHECK   To check your answer multiply:

$$2000$$
$$\times .09$$
$$\overline{180}$$

Here is another way to summarize how to get answers in the these three types of percent problems:

   In Type 1, you multiply the percent times the other number.

   In Type 2, you form a fraction, then divide.

   In Type 3, you set up an equation, then divide the other number by the percent.

Study these three types of percent problems. They will get you ready for the next section where you will practice solving word problems with percents. The exercises below review some basic concepts of percents and the 5-step strategy in preparation for the next section.

✎ *Exercises* _____

Identify each of the following as Type 1, 2 or 3. Do not work out the problem.

   1. 14 is what percent of 96? Type _____

   2. 350 is 25% of what number? Type _____

   3. What is 34% of 2350? Type _____

   4. If 23 is 55% of a number, what is the number? Type _____

   5. What percent of 240 is 36? Type _____

   6. Find 123% of 54. Type _____

Compute the answer to each of the following:

   7. What is 20% of 400? _____

   8. 16 is what percent of 64? _____

   9. 5 is 125% of what number? _____

   10. What percent of 20 is 8? _____

   11. 30 is what percent of 10? _____

   12. What is 40% of 125? _____

   13. What is the equivalent percent for $\frac{3}{10}$? _____

Choose the correct answer by circling the letter next to it.

14. What operation do we use to answer the question "What is 30 percent of 60"?
    a. addition
    b. subtraction
    c. multiplication
    d. division

15. In the HELP step of Type 2, what do we set up?
    a. an equation
    b. a fraction
    c. a diagram
    d. an addition example

16. In which step of the 5-step strategy, do we ask: is the answer reasonable?
    a. WHAT?
    b. HOW?
    c. HELP
    d. CHECK

17. In the HELP step of Type 3, what do we set up?
    a. an equation
    b. a fraction
    c. a diagram
    d. an addition example

18. In which step do we find the number of computation steps?
    a. WHAT?
    b. HOW?
    c. HELP
    d. COMPUTE

19. In which step do we figure out which operation to use?
    a. WHAT?
    b. HOW?
    c. HELP
    d. CHECK

20. In which step do we ask: what do we have to find in this problem?
    a. WHAT?
    b. HELP
    c. COMPUTE
    d. CHECK

6.3 How Many Computation Steps?

When you solve word problems, you have to answer the question, How many computation steps will it take to solve the problem?

You answer this question in the HOW step of the strategy. (Figuring out which type of percent problem is involved also belongs in the HOW step of the strategy.)

Many problems require only one computation step, while others require more than one computation step.

Here is an example that requires two computation steps:

EXAMPLE: TWO COMPUTATION STEPS

A stereo was on sale for 15% off the original price. The original price was $320. How much did the stereo sell for?

SOLUTION

Step 1. WHAT?   Find the price of the stereo.

Step 2. HOW?   Two things: First, Which type is this problem? It is Type 1. That means <u>multiply</u> to find 15% of 320. That is the first computation step.

The second computation step: subtract the answer to the first step from $320.

Step 3. HELP   Here is the setup for the two steps:

Step 1	Step 2
$320	Subtract the answer of
× 0.15	Step 1 from $320
?	

Step 4: COMPUTE Multiply and subtract:

Step 1	Step 2
$320	$320
× 0.15	− 48
$48	$272

The answer is $272.

Step 5. CHECK   Two parts:

$272
+ 48
$320

$$\frac{48}{320} = \frac{3}{20} = 0.15$$

$$0.15 = 15\%$$

✎ Exercises

For each of these problems find the number of computation steps required to find the answer. Do not work out the answers.

1. Muriel is a real estate agent. She receives 6% commission when she sells a house. She recently sold a house for $200,000. How much commission did she receive?

   Number of steps: _____

2. Mack works at the local hamburger place. Last Wednesday Mack made 125 hamburgers. There were 600 hamburgers made that day. What percent of all the hamburgers did Mack make?

   Number of steps: _____

3. Jodi bought a new business suit on sale at 20% off the original price of $80. She also bought shoes for $64. How much did she pay for her new clothes?

   Number of steps: _____

4. Two years ago 20% of Don's 200 hits were doubles. Last season 15% of his 180 hits were doubles. How many more doubles did he hit two seasons ago?

   Number of steps: _____

5. Karen bought a computer that was on sale for $165. The sale price was 75% of the original price. What was the original price?

   Number of steps: _____

6.4 Percents: Practice

Use the 5-step strategy to solve the 10 practice problems in this section. In the HOW step, determine which type the problem belongs to. Recognizing the type will help you decide which operation to use. Then find out how many computation steps it takes to solve the problem.

1. Jeff saw a VCR that was selling for $300. The salesperson at the store said that he could get the set at a 10% discount. How much does Jeff have to pay for the VCR?

Your answer: $ _____

2. Gale earned $1800 last summer. She paid 5% in income taxes on the first $1000 she earned and 15% on the rest of her income. How much money did she have left after taxes?

Your answer: $ _____

3. Jack bought $34 worth of goods at the supermarket. There was a tax of 6% on $5 of the bill. How much was the final total?

Your answer: $ _____

4. There are 30 students in Ned's night class. They are 10% of the total student enrollment at night school. How many students attend night school?

Your answer: _____

5. On a test, Maria was right 20 out of 25 times. What percent was she right?

Your answer: _____ %

6. Cindy received partial payment for the work that she did last week. She got 75% of the total amount she was owed. She got $150. How much was the total amount that she was owed?

Your answer: $ _____

7. Lunch was $8. A restaurant sales tax of 5% is added to the bill. How much change was there from $20?

Your answer: $ _____

8. Keith got on base 40% of the times he came to bat. And 75% of those times, he got a hit. If he came to bat 300 times, how many times did he get a hit?

Your answer: $ _____

9. A suit was reduced 25%. If the original price was $160, what was the selling price?

Your answer: $ _____

10. Alan received 55% of the vote for class treasurer. If 360 people voted, how many votes did Alan get?

Your answer: _____

6.5 *Percents: Checkout*

The purpose of this section is to give you a chance to check out your skills in solving percent word problems. Use the 5-step approach to get an answer and to check the answer.

For each example, you will find four answers. Select the correct answer for the problem by circling the letter next to the answer.

Sometimes, one of the choices will be NOT GIVEN. This choice means that the correct answer is not given among the other three choices. If you can't find the correct answer, then you should circle the letter next to NOT GIVEN.

1. Juan spends 20% of his take-home pay on food. His take-home pay is $600 per week. How much does he spend each week on food?

   a. $1200
   b. $600
   c. $120
   d. $100

2. The bill for food at the restaurant came to $25. The tax on this bill was $2. What percent is the tax on this bill?

   a. 2%
   b. 4%
   c. 6%
   d. NOT GIVEN

3. Jan bought dinner last night at the R & R Restaurant. The dinner cost $8.50. The tax was 6%. How much change was there from $20?

   a. $10.79
   b. $10.90
   c. $10.99
   d. $11.00

4. Lola received 40% of the votes in the general election. She received 440 votes. How many people voted?

   a. 264
   b. 176
   c. 1200
   d. 1100

5. Gloria bought a hat for $12 and a sweater for $34. If the sales tax is 5%, how much did she pay altogether?

   a. $44.10
   b. $46
   c. $48.30
   d. NOT GIVEN

6. Felix spends 30% of his leisure time watching TV. If he had 60 hours of leisure time last week, then how many hours did he watch TV last week?

   a. 12
   b. 16
   c. 18
   d. 20

7. Only 20% of the class, or 25 students, signed up for the trip. How many students were there in the class altogether?

   a. 5
   b. 15
   c. 100
   d. 125

8. Sue bought a jacket for $28. The tax was $1.40. What percent is the tax?

   a. 5%
   b. 10%
   c. 15%
   d. NOT GIVEN

9. Timmy did 30 pushups this morning. That is 60% of his goal. What is Timmy's goal?

   a. 18 pushups
   b. 50 pushups
   c. 55 pushups
   d. 90 pushups

10. Ginny bought a car for $6000. She paid $1000 of her own money and borrowed the rest of the money from the bank. The bank charged her 8% annual interest on the borrowed money. How much does she pay in interest each year?

   a. $200
   b. $400
   c. $600
   d. NOT GIVEN

7. TIME AND MONEY

7.1 Time and Money: Getting Started

Time and Money. You never have enough of them. (After this chapter you may feel that you have had too much of them!) With a small effort, you will be able to solve problems in these important areas. Time and money confront us every day, and there are many problems to solve.

The skills that are important in solving problems with time and money are these:

TIME: 1) Finding the difference between two times.
2) Finding equivalences between two times, such as the number of hours in a week.
MONEY: 3) Finding equivalences between two amounts of money, such as the number of nickels in $9.
4) Adding, subtracting, multiplying, and dividing amounts of money.

Here are some quick reminders of these skills that should help get you started in this chapter.

- -

Finding the difference between two times

To find the difference between a time in the morning and a time in the afternoon, use noon as a dividing time. Then figure the time before and after noon and add both times.

Example: How much time between 9 a.m. and 4 p.m.?

Before noon: 9 a.m. to 12—3 hours
After noon: 12 to 4 p.m.—4 hours
Total time = 7 hours

- -

Finding equivalences between two times

If you want to find how many smaller units are in a larger unit, multiply.

Example: How many minutes in a day?

Since there are 60 minutes in an hour and 24 hours in a day:

$$60 \times 24 = 1440 \text{ minutes}$$

- -

Finding equivalences between two amounts of money

Just as with time, if you want to find how many smaller units are in a larger unit, multiply.

Example: How many nickels in $7?

Since there are 20 nickels in 1 dollar:

$$20 \times 7 = 140 \text{ nickels}$$

Adding, subtracting, multiplying and dividing money

Use the same skills as with decimals. Don't forget to line up the decimal point.

Example: Subtract: $24.76 − $13.39 = ?

$$\begin{array}{r} \$24.76 \\ -\ 13.39 \\ \hline \$11.37 \end{array}$$

Use these exercises to review.

✎ Exercises

1. How many minutes are there in 300 seconds? _____

2. How many seconds are there in 45 minutes? _____

3. How many hours in 3 days? _____

4. How many weeks in two years? _____

5. How many days in 168 hours? _____

6. How many hours from 10:00 a.m. to 3:00 p.m. on the same day? _____

7. How many hours from 10:00 a.m. to midnight on the same day? _____

8. How many hours from 6:00 a.m. to 6:00 p.m. on the same day? _____

9. How many hours from 8:00 p.m. on Tuesday to 4:00 a.m. Wednesday? _____

10. How many hours from 2:00 p.m. Thursday to 6:00 p.m. Friday? _____

11. How many nickels in $2? _____

12. How many pennies in $5.59? _____

13. How many dollars in 3,400 pennies? _____

14. How many dollars in 1600 nickels? _____

15. How many dollars in 1600 dimes? _____

16. How many quarters in $16.25? _____

17. How many dollars in 464 quarters? _____

18. How many dimes in $36.70? _____

19. How many dimes in 400 nickels? _____

20. How many nickels in 500 quarters? _____

7.2 Time and Money: Examples

This section shows two word problems completely worked out—one on time and the other on money. Both examples use the 5-step strategy. If you are not sure about this strategy, look back to Chapter 1 where is was introduced. Here is a quick reminder about the five steps:

Step 1. WHAT?
Step 2. HOW?
Step 3. HELP
Step 4. COMPUTE
Step 5. CHECK

Remember the memory gem of Chapter 1:

WHAT! HOWARD HELPS COMPUTE CHECKS?

Each word in the sentence contains the name of one of the steps. Remember this sentence, and use the five steps to solve problems. Study each step in the following examples to see how it contributes to finding the answer. Don't forget the important last step—checking the answer.

- -

How Many Computational Steps

When you use the 5-step strategy, there is an important thing to remember in the HOW step. You must ask the question:

How many computation steps will it take to solve the problem?

Many problems require only one computation step. Sometimes you only have to subtract, or you only have to multiply to solve the problem. But some problems take more than one step. For example, you may have to subtract twice and then add to get the answer.

You need to recognize how may computation steps a problem requires. You do that in the HOW step. It is part of the strategy.

- -

EXAMPLE 1: Sam started a paint job at 8:30 a.m. and finished at 3:00 p.m. with a break of 45 minutes. How long did the paint job take?

SOLUTION:

Step 1. WHAT? To find the amount of time Sam worked.

Step 2. HOW? To find the amount of time use noon as a dividing line. Figure the time before and after noon and add them. Then change 45 minutes to part of an hour and subtract. That's 5 computation steps.

Step 3. HELP   Do the problem in steps:

1) Figure the amount of time before noon.
2) Figure the amount of time after noon.
3) Add steps 1 and 2.
4) Change 45 minutes to part of an hour.
5) Subtract step 4 from step 3.

Step 4. COMPUTE

(1) Before noon: 8:30 a.m. to 12— $3\frac{1}{2}$ hours

(2) After noon: 12 to 3 p.m.— $+3$ hours

$6\frac{1}{2}$ hours

(4) Change 45 minutes to $\frac{3}{4}$ hour

(5) $6\frac{1}{2} - \frac{3}{4} = 5\frac{3}{4}$

The answer is $5\frac{3}{4}$ hours, or 5 hours and 45 minutes

Step 5. CHECK   Add the answer and the break:

$5\frac{3}{4} + \frac{3}{4} = 6\frac{1}{2}$ hours

Start with 8:30 a.m. and add

8:30 to noon $= 3\frac{1}{2}$ hours

Three more hours makes it 3 p.m.

EXAMPLE 2: Jill went shopping at the supermarket. She bought meat for $10.54, fruit for $3.23 and bread for $2.86. How much change did she get from $20?

SOLUTION:

Step 1. WHAT?   To find the amount of change.

Step 2. HOW?   There are two steps. "How much change" suggests the second step is subtract.

Step 3. HELP   The setup for the two steps looks like this:

1) $10.54   2) Subtract Step 1
    3.23   from $20.
$+$ 2.86
     ?

Step 4. COMPUTE

1) $10.54   2) $20.00
    3.23 $-$ 16.63
$+$ 2.86 $3.37
$16.63

The answer is $3.37.

Step 5. CHECK

Start with the answer—$3.37.
Then add the 3 purchases:

$$
\begin{array}{rl}
\$ \quad 3.37 & \text{change} \\
10.54 & \text{meat} \\
3.23 & \text{fruit} \\
+\quad 2.86 & \text{bread} \\
\hline
\$20.00 & \text{money Jill started with}
\end{array}
$$

Go over these examples. They will help you to solve problems with money and time. The next section gives you further practice. The exercises below continue our review of some basic concepts of time and money in preparation for the next section.

✎ Exercises

1. How many hours in a week? _____

2. How many days in a year (not a leap year)? _____

3. How many minutes in a day? _____

4. How many seconds in 3 hours? _____

5. How many days in 11,520 minutes? _____

6. $23.45 − $16.69 = _____

7. $45.17 + $100.09 = _____

8. $2.65 × 3 = _____

9. $147.42 − $51.93 = _____

10. $45.36 ÷ 3 = _____

In the following problems choose the correct answer by circling the letter next to it.

11. In which step of the 5-step strategy do you decide how many computation steps there are?

    a. WHAT?
    b. HOW?
    c. HELP
    d. COMPUTE

12. In which step do we read the problem carefully?

    a. WHAT?
    b. HOW?
    c. HELP
    d. COMPUTE

13. In which step do we find the answer?

    a. WHAT?
    b. HOW?
    c. HELP
    d. COMPUTE

14. Which set of letters tells us the order of the five steps?

    a. H, H, C, C, W
    b. W, H, H, C, C
    c. C, C, H, H, W
    d. W, D, D, H, H

15. In which step do you decide what the question wants you to do?

    a. WHAT?
    b. HOW?
    c. HELP
    d. COMPUTE

16. In which step do you ask: is this answer reasonable?

    a. WHAT?
    b. HELP
    c. COMPUTE
    d. CHECK

17. In which step do you decide how to set up the computation?

    a. WHAT?
    b. HOW?
    c. HELP
    d. COMPUTE

7.3 Time and Money: Practice

Have you had enough of time and money? Just in case you haven't, here are 10 problems for you to practice. Use the 5-step procedure to solve the problems. Make sure to check each answer.

1. Tina's project is due tomorrow. She started working on her project at 8:15 a.m. and finished at 2:30 with 3/4 hour off for lunch. How long did Tina work on her project?

   Your answer: _____ hours _____ minutes

2. Joe works at the hardware store in town. He gets $4.50 per hour. How much does he earn after he works 24 hours?

   Your answer: $ _____

3. Barbara opened a new checking account with a deposit of $341.73. She then wrote checks for $42.35 and $15.87. At the end of the month the bank deducted $2.76 for a service fee. How much money did Barbara have left in her account?

   Your answer: $ _____

4. Manny loves to talk. He started his speech at 11:13 a.m. and finished at 12:25 p.m. How long did he talk?

   Your answer: _____ hours _____ minutes

5. Tony came home exhausted. The hiking trip he was on started on November 28 and ended on December 4. If you count the first and last days, how long was the trip?

   Your answer: _____ days

6. Ken loves to shop. He spent $34.12 for a sweater, $12.45 for a pair of gloves, and $8.93 for a book. If he started out with $70, how much did he have left after shopping?

   Your answer: $ _____

7. Mark has a part time job. He made $176.20 last week working five days. He earned the same amount each day. How much did he make each day?

   Your answer: $ _____

8. Ellen hates shopping, but she went anyway. She bought 3 pairs of socks for $2.15 each and 4 towels for $5.65 each. How much did she spend?

   Your answer: $ _____

9. Shelly went out for the evening. Dinner cost $14.85, movies $4 and the babysitter $7. How much did the evening cost?

   Your answer: $ _____

10. Dick had a class that lasted 200 minutes. How long is that in hours and minutes?
    Your answer: _____ hours _____ minutes

7.4 Time and Money: Checkout

The purpose of this section is to give you a chance to check out your skills in solving time and money word problems.

For each example, you will find four answers. Select the correct answer for the problem by circling the letter next to the answer.

Sometimes, one of the choices will be NOT GIVEN. This choice means that the correct answer is not given among the other three choices. If you can't find the correct answer, then you should circle the letter next to NOT GIVEN.

1. Corry loves TV. He watches TV from 4:00 p.m. until 7:30 p.m. with a short break of 20 minutes. How long does Corry watch TV?

    a. 2 hours and 20 minutes
    b. 3 hours 10 minutes
    c. 3 hours 20 minutes
    d. NOT GIVEN

2. The gang went out for pizza after school. Joan had a regular pizza ($1.25), Phil had a mushroom pizza ($1.45), and Terry had anchovies on her pizza ($1.75). Drinks came to $1.80. How much was the total bill?

    a. $6.00
    b. $6.25
    c. $7.25
    d. NOT GIVEN

3. A train took 5 hours and 23 minutes to travel from Shreveport to New Orleans. The train stopped four times for 5 minutes each, and there was one 12-minute stop to pick up passengers. What was the actual travel time of the train?

    a. 4 hours 51 minutes
    b. 4 hours 41 minutes
    c. 4 hours 31 minutes
    d. 4 hours 30 minutes

4. Lola works in her jewelry studio from 2:30 p.m. until 7:30 p.m. on Monday, Tuesday, and Wednesday. On Thursday and Friday she works from 2:00 p.m. to 8:30 p.m. What is the total amount of time that Lola works in the studio?

    a. 22 hours
    b. 24 hours
    c. 26 hours
    d. 28 hours

5. Steve is state champion in tennis. Yesterday he bought a tennis racket for $125 and sneakers for $65. He paid with a check for $250. How much money did he get back?

    a. $30
    b. $50
    c. $55
    d. NOT GIVEN

6. Boris went to bed at 9:45 p.m. and woke up at 7:15 a.m. How long did he sleep?

    a. 8 hours 15 minutes
    b. 8 hours 30 minutes
    c. 9 hours 15 minutes
    d. 9 hours 30 minutes

7. The special TV movie ran was scheduled from 8 p.m. until 10 p.m. It had eight 30-second commercials and four 60-second commercials. How long was the actual movie, not counting commercials?

    a. 1 hour 56 minutes
    b. 1 hour 54 minutes
    c. 1 hour 52 minutes
    d. 1 hour 50 minutes

8. Carla took $25 and spent it on the following: $13.24 for coffee, $2.38 for sugar, and $1.41 for milk. How much money did she have left?

    a. $7.97
    b. $8.07
    c. $8.27
    d. NOT GIVEN

9. Denise loves cars. She saved $3500 so she can buy a car. She saw one she liked, but it costs $5750. Then she found out that registration would cost $25 and the state tax would be $345. How much more money does she need?

    a. $2620
    b. $2720
    c. $3620
    d. $3820

10. Gary loves to eat. He spent $3.45 on hamburgers, $2.50 on ice cream, and $1.75 on drinks. He got 30 cents change. How much money did he pay with?

    a. $10
    b. $9
    c. $8
    d. $7

11. One morning Steve calculated how much time he spent getting to work. He calculated that he walked for 12 minutes, was on a train for 37 minutes, and on a bus for 13 minutes. His waiting time was 14 minutes. How much time did he spend getting to work?

    a. 1 hour 6 minutes
    b. 1 hour 10 minutes
    c. 1 hour 16 minutes
    d. 1 hour 20 minutes

8. *Ratios and Proportions*

8.1 Ratios and Proportions: Getting Started

This chapter is about ratios and proportions.

A **ratio** is the comparison of two numbers. We usually write ratios with as a fraction like this:

$$\frac{3}{7} \text{ (read as "3 is to 7").}$$

Sometimes you will see a ratio written with a colon:

$$3:7 \text{ (also read as "3 is to 7").}$$

The ratio $\frac{3}{7}$ compares 3 to 7. It might be comparing wins to losses, boys to girls, trains to planes, or apples to oranges. In other words, a ratio can be used to compare any two quantities.

A **proportion** is a statement that two ratios are equal. Here is an example:

$$\frac{3}{7} = \frac{6}{14}$$

This proportion means, for example, that the ratio of 3 oranges for every 7 apples is the same as 6 oranges for every 14 apples. You read this proportion as "3 is to 7 as 6 is to 14."

In solving word problems you will frequently set up a proportion that has a letter in place of a number. This letter is called a **variable**. Here is an example:

$$\frac{n}{15} = \frac{2}{5}$$

To find n, use the technique of **cross multiplication:**

$$\frac{n}{15} = \frac{2}{5}$$

$$n \times 5 = 2 \times 15$$

$$n \times 5 = 30$$

Divide both sides by 5:

$$n = 6$$

Study this method, because you will need it often.

The problems in this chapter are all about ratios and proportions. In case you have forgotten about ratios and proportions, the exercises below remind you about them. If you need more help, review these topics further before you begin this chapter.

Write a ratio with two whole numbers for each of these statements.

1. There were 17 days of rain and 13 days of sunshine in June. _____

2. The Broncos won 11 games and lost 9. _____

3. There were 14 girls and 10 boys in class. _____

4. Jennifer read 136 pages and had 101 pages to go. _____

5. There are 60 minutes in 1 hour. _____

6. Alberto drove 90 miles in 3 hours. _____

7. One quart of orange juice sells for 97 cents. _____

8. A can of twelve ounces of tuna sells for $3.00. _____

9. Fred stuffed 10 envelopes in $\frac{1}{2}$ hour. _____

10. Nancy painted $3\frac{1}{2}$ chairs in 2 hours. _____

In the next 10 problems, there is one number missing in each proportion. The missing number is shown by a letter—a variable. Find the missing number.

11. $\dfrac{2}{3} = \dfrac{4}{s}$ _____

12. $\dfrac{2}{5} = \dfrac{t}{45}$ _____

13. $\dfrac{x}{3} = \dfrac{80}{5}$ _____

14. $\dfrac{32}{5} = \dfrac{b}{10}$ _____

15. $\dfrac{q}{42} = \dfrac{3}{7}$ _____

16. $\dfrac{1}{5} = \dfrac{5}{c}$ _____

17. $\dfrac{200}{500} = \dfrac{800}{w}$ _____

18. $\dfrac{a}{10} = \dfrac{12}{30}$ _____

19. $\dfrac{7}{g} = \dfrac{14}{6}$ _____

20. $\dfrac{100}{800} = \dfrac{8}{p}$ _____

Here are two word problems on ratio and proportion completely worked out. Both examples use the 5-step strategy introduced in Chapter 1.

If you are not sure about this strategy, look back to Chapter 1. These are the five steps that make problem solving easier for you:

Step 1. WHAT?
Step 2. HOW?
Step 3. HELP
Step 4. COMPUTE
Step 5. CHECK

Did you forget the memory gem of Chapter 1:

WHAT! HOWARD HELPS COMPUTE CHECKS?

Each word in the sentence contains the name of one of the steps. Try to remember this sentence, and use the five steps to solve problems. Study each step in the following examples to see how it contributes to finding the answer. Don't forget the important last step—checking the answer.

- -

How Many Computation Steps

When you use the 5-step strategy, there is an important thing to remember in the HOW step. You must ask the question:

How many computation steps will it take to solve the problem?

Many problems require only one computation step. Sometimes you only have to subtract, or you only have to multiply to solve the problem. But some problems take more than one step. For example, you may have to multiply and then add to get the answer.

You need to recognize how many computation steps a problem requires. You do that in the HOW step. It is part of the strategy.

- -

EXAMPLE 1: Billy's car traveled 160 miles on 9 gallons of gasoline. How many gallons does Billy's car need to travel 640 miles?

SOLUTION:

Step 1. WHAT?   To find how many gallons of gas needed for 640 miles.

Step 2. HOW?   Use a proportion. One computation step.

Step 3. HELP   Set up this proportion:

$$\frac{160}{9} = \frac{640}{n}$$

Step 4. COMPUTE Cross multiply: $160 \times n = 9 \times 640$

$$160 \times n = 5760$$

$$n = 36$$

The answer is 36 gallons.

Step 5. CHECK   Write the proportion again this time with the answer:

$$\frac{160}{9} = \frac{640}{36}$$

$$160 \times 36 = 9 \times 640$$

$$5760 = 5760$$

- -

EXAMPLE 2: The ratio of boys to girls who went out for sports last year was 6 to 5. If 143 students went out for sports, then how many were girls?

Note: Example 2 is tricky. Here's why. You are given the <u>ratio</u> of boys to girls, but you are not given the <u>number</u> of either one. Instead, you are given the <u>total number of students</u> (boys plus girls). You can't make a proportion unless you first change part of the ratio. The HOW and HELP steps will show you what to do.

SOLUTION:

Step 1. WHAT?   To find out how many girls went out for sports.

Step 2. HOW?   First find the ratio of girls to the total number of students. Then use a proportion. Three computation steps.

Step 3. HELP   1) Add: $6 + 5 = 11$ (the total number of students).
2) The ratio of girls to total number of students is 5 to 11.
3) Set up this proportion:

$$\frac{5}{11} = \frac{n}{143}$$

Step 4. COMPUTE Cross multiply:

$$11 \times n = 5 \times 143$$

$$11 \times n = 715$$

$$n = 65$$

The number of girls is 65.

Step 5. CHECK   Replace n with 65 in the proportion:

$$\frac{5}{11} = \frac{65}{143}$$

Cross multiply:

$$5 \times 143 = 11 \times 65$$

$$715 = 715$$

90

Study these two examples. They will get you ready for the next section. In the next section, you will practice solving word problems with ratios and proportions. The exercises below review some more basic concepts of ratios and proportions in preparation for the next section.

✎ Exercises

How do you read each of these ratios?

1. 5:6 _____(See Section 8.1 if you've forgotten.)

2. 23:59 _____

3. $\dfrac{54}{33}$ _____

4. $\dfrac{100}{50}$ _____

How do you read these proportions?

5. $\dfrac{2}{3} = \dfrac{4}{6}$ _____

6. $\dfrac{1}{2} = \dfrac{5}{10}$ _____

7. 9:3 = 3:1 _____

8. 100:1 = 500:5 _____

For Exercises 9–14 a letter stands for a missing number in each proportion. Find the missing number.

9. $\dfrac{25}{4} = \dfrac{125}{a}$ _____

10. $\dfrac{13}{b} = \dfrac{52}{8}$ _____

11. $\dfrac{8}{3} = \dfrac{160}{c}$ _____

12. $\dfrac{2}{7} = \dfrac{d}{91}$ _____

13. $\dfrac{2100}{9} = \dfrac{700}{f}$ _____

Choose the equivalent ratio. Circle the correct answer.

14. $\dfrac{7}{8} =$

   a. $\dfrac{21}{32}$ **b.** $\dfrac{28}{21}$ **c.** $\dfrac{35}{40}$ **d.** $\dfrac{35}{32}$

15. $\dfrac{3}{4} =$

   a. $\dfrac{21}{28}$ **b.** $\dfrac{6}{9}$ **c.** $\dfrac{21}{24}$ **d.** $\dfrac{12}{20}$

16. $\dfrac{2}{3} =$

   a. $\dfrac{18}{24}$ **b.** $\dfrac{24}{33}$ **c.** $\dfrac{16}{21}$ **d.** $\dfrac{22}{33}$

17. $\dfrac{5}{4} =$

   a. $\dfrac{28}{35}$ **b.** $\dfrac{40}{44}$ **c.** $\dfrac{45}{36}$ **d.** $\dfrac{50}{45}$

18. $\dfrac{7}{8} =$

   a. $\dfrac{14}{12}$ **b.** $\dfrac{63}{72}$ **c.** $\dfrac{35}{45}$ **d.** $\dfrac{35}{32}$

Choose the correct answer by circling the letter next to it.

19. In which step of the 5-step strategy, do we ask: is the answer reasonable?

   a. WHAT?
   b. HOW?
   c. HELP
   d. CHECK

20. In which step do we determine the number of computation steps?

   a. WHAT?
   b. HOW?
   c. HELP
   d. CHECK

21. In which step do we find the answer?

    a. WHAT?
    b. HOW?
    c. HELP
    d. COMPUTE

22. In which step do we figure out which operation to use?

    a. WHAT?
    b. HOW?
    c. HELP
    d. CHECK

23. In which step do you read the problem carefully to find what the question wants you to do?

    a. WHAT?
    b. HOW?
    c. HELP
    d. COMPUTE

8.3 Ratios and Proportions: Practice

In this section you will find 10 problems to practice. Use the 5-step procedure to solve the problems, and make sure to check each answer.

1. Alma drove 100 miles in 2 hours. At that rate, how long will it take her to drive 250 miles?

   Your answer: _____ hours

2. Dan works at a pizza shop. For every 3 cheese pizzas that Dan makes, he makes 1 mushroom pizza. If he made 12 mushroom pizzas, how many cheese pizzas did he make?

   Your answer: _____

3. Mac works in a fast-food restaurant. He sells 3 hamburgers for every 1 hot dog he sells. On Friday he sold 150 hamburgers. How many hot dogs did he sell?

   Your answer: _____

4. Jan can type 4 pages every 12 minutes. At that rate, how many pages can she type in 3 hours?

   Your answer: _____

5. The number of companies whose stock dropped in value compared to the number that rose was 7 to 5. If 1200 stocks traded, how many went up in value?

   Your answer: _____

6. According to the school paper, there are 10 students for every 3 teachers. At that ratio, how many teachers are there if there are 420 students?

   Your answer: _____

7. Vanessa is baking a cake. The recipe requires 2 cups of milk for every 5 cups of flour. She uses 11 cups of milk. How many cups of flour does she need?

   Your answer: _____

8. Maria drove 325 miles and used 13 gallons of gasoline. At that rate, how much gasoline will her car use if she drives 625 miles?

   Your answer: _____ gallons

9. Carmen works in a supermarket. Carmen said that the ratio of apples to oranges is 7 to 4. There are 1100 pieces of fruit altogether. How many are apples?

   Your answer: _____

10. Kitty is getting ready for a party. For the party Kitty figures that she will need 2 pizzas for every 8 guests. How many pizzas will she need for 40 guests?

    Your answer: _____

8.4 Ratios and Proportions: Checkout

The purpose of this section is to give you a chance to check out your skills in solving ratio and proportion word problems.

For each example, you will find four answers. Select the correct answer for the problem by circling the letter next to the answer.

For some problems, one of the choices is NOT GIVEN. This choice means that the correct answer is not given among the other three choices. If you can't find the correct answer, then you should circle the letter next to NOT GIVEN.

1. A drug store's records from last year show that the store sold 3 bottles of aspirin for every 2 bottles of shampoo. The records show that the store sold 2854 bottles of shampoo last year. How many bottles of aspirin did the store sell?

   a. 8562

   b. 1904

   c. 4281

   d. NOT GIVEN

2. It was the first day of football practice. The ratio of running backs to linesmen at the practice was 5 to 4. If 40 running backs showed up, how many linesmen came to practice?

   a. 60

   b. 50

   c. 32

   d. NOT GIVEN

3. Aliza was having a party. She figured she needs 2 quarts of punch for every 7 guests. She made 8 quarts of punch. How many guests was Aliza expecting?

   a. 26

   b. 27

   c. 28

   d. 30

4. Chang does crossword puzzles. For every 2 that he can finish, there are 25 he can't finish. If he tries 324 puzzles how many can he finish?

    a. 22

    b. 25

    c. 27

    d. NOT GIVEN

5. Kristina works at a swimming club. She says that for every 6 adults who come to the club there are 4 children. How many children come to the club when the total number of people is 90?

    a. 24

    b. 36

    c. 60

    d. NOT GIVEN

6. Tina works in a dessert shop. In a recent survey, the shop found that for every 5 people who buy maple walnut ice cream, 12 people buy some other flavor. How many people buy maple walnut ice cream if 156 do not?

    a. 60

    b. 65

    c. 70

    d. 75

7. Jeff works for a newspaper. He says that for every 4 paragraphs he reads he finds 10 mistakes. How many mistakes does he find when he reads 40 paragraphs?

    a. 40

    b. 60

    c. 80

    d. 100

8. There are 5 new computers for every 9 old computers. How many new ones are there if there are 56 computers altogether?

    a. 18

    b. 22

    c. 26

    d. NOT GIVEN

9. Fred works in a diner. He says that for every 10 customers, 4 order coffee. If Fred is right, how many customers drink coffee when there are 55 customers?

    a. 11

    b. 14

    c. 22

    d. NOT GIVEN

10. At the tennis camp the ratio of girls to boys was 6 to 5. If there were 121 people altogether, how many were girls?

    a. 55

    b. 66

    c. 70

    d. 110

9. *Review Test B*

This test covers the topics of Chapters 6, 7 and 8. It consists of 20 questions covering all three topics— 1) percents, 2) time and money, and 3) ratio and proportion.

There are four answers for each example. Select the correct answer to the problem by circling the letter next to the answer.

For some of the problems, one of the choices is NOT GIVEN. This choice means that the correct answer is not given among the other three choices. If you can't find the correct answer, then circle the letter next to NOT GIVEN.

1. Forty percent of the freshmen class came to the basketball game. If there are 230 freshmen, how many came to the basketball game?

   a. 82

   b. 92

   c. 120

   d. NOT GIVEN

2. Ann started typing her report at 5:00 p.m. on Tuesday. She took a break at 7:30 p.m. for 1½ hours. Then she continued typing until 11:15 p.m. How long did she work on her report?

   a. 2 hours 15 minutes

   b. 2 hours 45 minutes

   c. 3 hours 15 minutes

   d. 4 hours 45 minutes

3. Carla works in a diner. She says that the ratio of people who drink coffee to those who drink tea is 7:1. If there are 140 people who drink coffee, then how many people drink tea?

   a. 20

   b. 40

   c. 120

   d. 160

4. The ratio of cloudy days to sunny days in June is 2:13. How many cloudy days would you expect in this month?

   a. 1 day

   b. 2 days

   c. 3 days

   d. 4 days

5. Jack works in the bank as a teller. A customer came in Friday with a $100 bill, and asked for quarters. How many quarters did Jack give the customer for the $100?

    a. 100

    b. 200

    c. 300

    d. 400

6. Terry bought a coat that was on sale. The original price was $80. It was on sale for 20% off the original price. How much did Terry pay for the coat?

    a. $16

    b. $54

    c. $60

    d. NOT GIVEN

7. Karen drives to work every day. It takes her 37 minutes to get to work and 43 minutes to get home. If she works five days each week, how long does she spend commuting to and from work?

    a. 6 hours 45 minutes

    b. 6 hours 40 minutes

    c. 6 hours 30 minutes

    d. NOT GIVEN

8. Sam said that for every 3 people who eat fish in the museum cafeteria, 5 people eat chicken. If 175 people eat chicken, then how many people eat fish?

    a. 95

    b. 105

    c. 125

    d. 420

9. Tanya could have danced all night. She danced from 6:30 p.m. to 9:15 p.m., then rested. She started dancing again from 10:15 until 12 midnight. How long did she dance altogether?

   a. 3½ hours

   b. 4 hours

   c. 4½ hours

   d. 5 hours

10. Elie's doctor's bills last month were high. She paid her doctor five times: $25, $65, $70, $35 and $40. She got $125 back from her health insurance company. How much did her doctor cost her last month?

   a. $360

   b. $235

   c. $110

   d. $100

11. Gilda was unhappy. Only 48% of the people she invited came to her party. She invited 25 people. How many people came to Gilda's party?

   a. 10

   b. 12

   c. 13

   d. 16

12. Orrie says that the ratio of men to women taking science courses is 5:3. If 126 women take science courses, then how many men take science courses?

   a. 210

   b. 220

   c. 300

   d. 420

13. Peggy spends 30% of her monthly income on food. Peggy's monthly income is $1260. How much does she spend on food?

    a. $368

    b. $388

    c. $478

    d. NOT GIVEN

14. Don went to the bank with a bag of nickels. He had $40 worth of nickels. How many nickels did he have?

    a. 400

    b. 800

    c. 1000

    d. 1600

15. In a recent poll, 80% of the athletes at school do not eat breakfast. There are 110 athletes at school. How many athletes do not eat breakfast?

    a. 44

    b. 66

    c. 80

    d. NOT GIVEN

16. Sally answers the phone 40% of the time in her house. If the phone rings 300 times a month, how many times does she answer the phone?

    a. 120

    b. 130

    c. 140

    d. 150

17. Lisa works very hard. Whenever she works over 40 hours she gets 1.5 times her regular hour wage. She gets paid $6.00 per hour. How much does she gross each week when she works 60 hours a week?

   a. $240

   b. $360

   c. $420

   d. NOT GIVEN

18. Ed saves 10% of his gross earnings. He earns $7 per hour, and works 35 hours a week. How much does he save each week?

   a. $17.50

   b. $19

   c. $24

   d. NOT GIVEN

19. Light travels at the rate of 186,000 miles per second. How far does it travel in a minute?

   a. 11,160,000 miles   11 160 000

   b. 1,116,000 miles   1 116 000

   c. 31,000 miles   31 000

   d. 3100 miles   3 100

20. Ed counted the different letters in several pages of the book he was reading. He found that the ratio of e's to s's was 5:2. He found 75 e's. How many s's did he find?

   a. 20

   b. 30

   c. 40

   d. 50

10. *Measurement*

10.1 Measurement: Getting Started

This chapter is about measurement. We use measurement skills whenever we weigh something, or find out how long an object is, or figure out the capacity of a container.

Think about measurement. What do you think about—feet and inches, pounds and ounces, pints and quarts? If so, you're right. These units are a good part of measurement, but so are degrees Fahrenheit and degrees Celius and meters, liters, and grams.

Here are two main things you have to know when it comes to solving problems in measurement:

1. **Units of measurement**—the units that measure length, weight, capacity and temperature.

2. **Measurement equivalents**—the number of inches in a foot, the number of centimeters in a meter, the number of quarts in a gallon, and so on.

There are some abbreviations that you should know about as you work through this chapter. Notice that the only abbreviations using capital letters are those for temperature, and that the only abbreviation using a period is the one for inch. (This is the most modern way of writing abbreviations. Older books used periods after many of these abbreviations.)

TABLE OF ABBREVIATIONS		
	CUSTOMARY UNITS	METRIC UNITS
Length	in. = inch ft = feet yd = yard	mm = millimeter cm = centimeter m = meter km = kilometer
Weight	oz = ounce lb = pound t = ton	mg = milligram g = gram kg = kilogram
Capacity	oz = ounce qt = quart gal = gallon	ml = milliliter l = liter kl = kiloliter
Temperature	°F = degrees Fahrenheit	°C = degrees Celsius (formerly Centigrade)

TABLE OF EQUIVALENCES

	CUSTOMARY UNITS	METRIC UNITS
Length	12 in. = 1 ft 3 ft = 1 yd 1760 yd = 1 mile	10 mm = 1 cm 100 cm = 1 m 1000 m = 1 km
Weight	16 oz = 1 lb 2000 lb = 1 t	1000 mg = 1 gm 1000 g = 1 kg
Capacity	2 cups = 1 pt 2 pt = 1 qt 4 qt = 1 gal	1000 ml = 1 l 1000 l = 1 kl

Converting from one measure to another

To convert from one measure to another you either multiply or divide.

Example: How many inches in 14 feet?

There are 12 inches in 1 foot. Since the inch is the smaller unit, we multiply by 12:

$$12 \times 14 = 168 \text{ in.}$$

Example: How many meters in 783 centimeters?

There are 100 centimeters in 1 meter. Since the meter is the larger unit, we divide by 100:

$$783 \div 100 = 7.83 \text{ m}$$

The problems in this chapter are all about measurement. In case you forgot about the different types of units, the exercises below will remind you. If you need more help, review this topic further before you begin this chapter.

✏️ Exercises

Circle the letter next to the best unit of measure for each.

1. Length of a room

    a. tons
    b. quarts
    c. feet
    d. grams

2. Amount of juice in a container

    a. meters
    b. degrees
    c. quarts
    d. feet

3. Temperature this morning

    a. degrees Fahrenheit
    b. pints
    c. inches
    d. liters

4. Weight of a box of cereal

    a. liters
    b. feet
    c. grams
    d. kilometers

5. Length of a book

    a. grams
    b. centimeters
    c. quarts
    d. yards

6. Weight of a car

    a. ounces
    b. feet
    c. tons
    d. yards

7. Recipe for pudding for 3 people

    a. meter
    b. yard
    c. cup
    d. gallon

Change each measure to the equivalent units indicated in each exercise.

8. 36 oz = _____ lb _____ oz

9. 43 in. = _____ ft _____ in.

10. 13 qt = _____ gal _____ qt

11. 249 cm = _____ m _____ cm

12. 5 kl = _____ l

13. 7,681 m = _____ km _____ m

14. 4 lb 7 oz = _____ oz

15. 47 ft = _____ yd _____ ft

16. 13 pt = _____ qt _____ pt

17. 7 gal = _____ qt

18. 2 yd = _____ in.

19. 3 t = _____ lb

20. 7 km = _____ m

21. 8,560 cm = _____ m _____ cm

22. 19 kg = _____ g

23. 3,000 mg = _____ g

24. 8 cups = _____ pt

25. 4 m = _____ mm

26. 17 l = _____ ml

27. 5 miles = _____ yd

10.2 *Measurement: Examples*

In this section you will find two examples of word problems that involve measurements. Both are completely worked out. The solutions for both examples follow the five-step strategy of Chapter 1.

These are the five steps that make problem solving easier for you:

> **Step 1.** WHAT?
> **Step 2.** HOW?
> **Step 3.** HELP
> **Step 4.** COMPUTE
> **Step 5.** CHECK

Remember the memory gem of Chapter 1:

> WHAT! HOWARD HELPS COMPUTE CHECKS?

Each word in this sentence contains the name of one of the steps. Try to remember this sentence, and use the five steps to solve problems.

Study each step of the two examples below to see how it contributes to finding the answer.

Don't forget: In the HOW step of the strategy you have to answer the question:

How many computation steps will it take to solve the problem?

Many problems require only one computation step. These are called 1-step problems. Other problems require more than one computation step. These are called 2-step, 3-step, etc. depending upon the number of computation steps it takes. Example 1 is a 1-step problem.

EXAMPLE 1   Carl and Zubin entered a football-throwing contest. Carl threw a football 33 meters 47 centimeters. Zubin threw a football 35 meters 18 centimeters. How much farther did Zubin throw a football than Carl?

SOLUTION

Step 1. WHAT?   How much farther did Zubin throw a football than Carl?

Step 2. HOW?   Subtract. There is one computation step.

Step 3. HELP   Here's the setup:

$$\begin{array}{r} 35 \text{ m } 18 \text{ cm} \\ -33 \text{ m } 47 \text{ cm} \\ \hline ? \end{array}$$

Step 4. COMPUTE $\begin{array}{r} 35 \text{ m } 18 \text{ cm} \\ -33 \text{ m } 47 \text{ cm} \end{array} = \begin{array}{r} 34 \text{ m } 118 \text{ cm} \\ 33 \text{ m } 47 \text{ cm} \\ \hline 1 \text{ m } 71 \text{ cm} \end{array}$

The answer is 1 m 71 cm.

Step 5. CHECK   Add:

$$
\begin{array}{r}
1\ \text{m}\quad 71\ \text{cm} \\
+\ \ 33\ \text{m}\quad 47\ \text{cm} \\
\hline
34\ \text{m}\ 118\ \text{cm}\ =\ 35\ \text{m}\ 18\ \text{cm}
\end{array}
$$

- -

EXAMPLE 2   Luis went to the store to pick up six packages of paper. Three packages weighed 14 ounces and three packages weighed 12 ounces. How much did the six packages weigh altogether?

SOLUTION

Step 1. WHAT?   To find the weight of the 6 packages.

Step 2. HOW?   Multiply twice and add. There are three computation steps.

Step 3. HELP   Here is the setup:

Step 1	Step 2	Step 3
14 oz	12 oz	Add the answers
× 3	× 3	of Steps 1 and 2
?	?	

Step 4. COMPUTE

Step 1	Step 2	Step 3
14 oz	12 oz	42 oz
× 3	× 3	+ 36 oz
42 oz	36 oz	78 oz

The answer is 78 oz, or 4 lb 14 oz (Changing 78 oz to 4 lb 14 oz might count as another step. You have to divide 78 oz by 16. Your answer comes to 4 remainder 14, or 4 lb 14 oz.)

Step 5. CHECK   Ask: Is the answer reasonable?
To check: divide twice: $36 \div 3 = 12$
$42 \div 3 = 14$

- -

Study these two examples. They will help prepare you for the next section. In the next section, you will practice solving word problems with measurement. The exercises below review some additional basic computations with measurements in preparation for the next section.

✎ Exercises

Do the following computations. Be sure you change any answers like 2 ft 14 in. to the form 3 ft 2 in. (You will have to do this in nearly every problem.)

1.   12 ft 10 in.
          + 10 ft 9 in.

2.   45 yd 1 ft
          − 18 yd 2 ft

3.   5 lb 12 oz
              × 7
   _____

4.   34 m 68 cm
      + 12 m 95 cm
   _____

5.   81 m 42 cm
      − 15 m 67 cm
   _____

6.   42 qt 0 pt
       − 28 qt 1 pt
   _____

7.   81 kg 245 g
      − 29 kg 368 g
   _____

8.   34 kl 429 l
       + 23 kl 982 l
   _____

9.   6 gal 2 qt
       − 2 gal 3 qt
   _____

10.   4 t 1452 lb
       +6 t 1982 lb
   _____

11.   14 km 498 m
       − 9 km 891 m
   _____

12.   37 cm 8 mm
       + 22 cm 5 mm
   _____

13.   17 m 23 cm
               × 5
   _____

14.   14 ft 6 in.
                × 9
   _____

15.   10 pt 1 cup
                × 4
   _____

16.   2 kl 891 l
         − 1 kl 900 l
   _____

Find the equivalent measure for each of these.

17. 32 oz = _____ lb

18. 3 miles = _____ yd

19. 3/4 lb = _____ oz

20. 3 gal = _____ qt

21. 1/2 gal = _____ qt

22. 8 qt = _____ gal

23. 72 in = _____ ft

24. 20 t = _____ lb

25. 1 mile = _____ ft

26. 30 mm = _____ cm

27.   In which step of the 5-step strategy do we ask, "Is the answer reasonable?"

   a. WHAT?
   b. HOW?
   c. HELP
   d. CHECK

28. In which step do we read the problem carefully?

    a. WHAT?
    b. HOW?
    c. HELP
    d. CHECK

29. In which step do we find the answer?

    a. WHAT?
    b. HOW?
    c. HELP
    d. COMPUTE

30. In which step do we decide on which operation to use?

    a. WHAT?
    b. HOW?
    c. HELP
    d. CHECK

10.3 Measurement: Practice

In this section you will find 10 problems to practice. Use the 5-step procedure to solve the problems, and make sure to check each answer. Make sure your answers are not in a form like 1 lb 17 oz. An answer like this should be converted to 2 lb 1 oz.

1. Seth weighed three packages. They weigh 2 lb 12 oz, 4 lb 7 oz, and 6 lb 14 oz. How much do they weigh altogether?
   Your answer: _____ lb _____ oz

2. Jodi's family has a car and a truck. The car weighs 2 t 560 lb and the truck weighs 4 t 150 lb. How much more does the truck weigh?
   Your answer: _____ t _____ lb

3. Aliza is planning a party. She wants to make 16 quarts of punch. How many gallons of punch does she want to make?
   Your answer: _____ gal

4. Mount McKinley in Alaska is about 6.2 km high. About how many meters high is Mount McKinley?
   Your answer: _____ m

5. The average annual rainfall in Pago Pago, Samoa, is 16.13 ft. What is the average annual rainfall in inches?
   Your answer: _____ in.

6. Kal works in a computer store. The new microcomputers weigh 24 lb 12 oz each. How much do 5 of these microcomputers weigh?
   Your answer: _____ lb _____ oz

7. Judah bought his favorite cereal at the supermarket. He bought 6 boxes. Each box weighs 453 g. How much do these 6 boxes weigh altogether?
   Your answer: _____ kg _____ g

8. Cynthia followed a recipe carefully. She put 19 cups of water into the big pot. How many gallons of water did she put into the water?
   Your answer: _____ gal _____ pt _____ cup

9. Dwight drives his car 18 km to work each day. How many meters does he drive to work?
   Your answer: _____ m

10. Ted works in a chemistry lab. He needs to fill 50 glasses with water. Each glass holds 140 ml of water. How many liters of water will he use?
    Your answer: _____ l

The purpose of this section is to give you a chance to check out your skill in solving measurement problems.

There are four answers for each example. Select the correct answer to the problem by circling the letter next to the answer.

For some of the problems, one of the choices is NOT GIVEN. This choice means that the correct answer is not given among the other three choices. If you can't find the correct answer, then circle the letter next to NOT GIVEN.

Tables of abbreviations and equivalences are available in Section 10.1.

1. The players on the Hawks use a bat 36 in. long. How many feet long is each bat?

   a. 2 ft.
   b. 3 ft.
   c. 12 ft.
   d. NOT GIVEN

2. A baseball company shipped 20,000 boxes of baseballs weighing 7 oz. per box. How many tons were shipped?

   a. 20 t
   b. 70 t
   c. 140 t
   d. NOT GIVEN

3. A package weighs 5 kg. How many grams does it weigh?

   a. 2000 g
   b. 3000 g
   c. 10,000 g
   d. NOT GIVEN

4. The orange can of orange juice holds 3 pt. The yellow can of orange juice holds 1 qt. Which can holds more juice? How much more?

   a. yellow—1 pt
   b. orange—1 pt
   c. yellow—2 pt
   d. orange—2 pt

5. Choose the most reasonable temperature for a cool fall day.

   a. 20 degrees F
   b. 50 degrees F
   c. 95 degrees F
   d. 105 degrees F

6. On a baseball diamond, home plate is 66 ft 6 in. from the pitcher's mound. How many inches is this distance?

   a. 666 in.
   b. 792 in.
   c. 802 in.
   d. NOT GIVEN

7. Tom measured his thumb. It is 5 cm 4 mm long. How many mm is this?

   a. 9 mm
   b. 45 mm
   c. 54 mm
   d. 504 mm

8. Choose the best estimate for the capacity of a tablespoon.

   a. 12 ml
   b. 120 ml
   c. 1.2 l
   d. 12 l

9. A bottle of apple juice holds 1.7 l. How many ml of juice is this?

   a. 1.7 ml
   b. 17 ml
   c. 170 ml
   d. NOT GIVEN

10. A box of raisins weigh 15 grams. If there are 30 raisins in a box, about how much does each raisin weigh?

    a. 2 g
    b. 4 g
    c. 1.5 g
    d. 0.5 g

11. *Perimeter, Area, and Volume*

11.1 Perimeter, Area, and Volume: Getting Started

This chapter contains problems on perimeter, area and volume. There are two main factors that will help you solve these kinds of problems:

1. Recognizing the shapes of different geometric figures and
2. Knowing the formulas for perimeter, area, and volume.

In case you have forgotten the different shapes and the formulas, the tables below will help remind you. Look at it carefully before you continue. You will be referring back to it often as you work through this chapter.

1. AREA (A)			
shape	**diagram**	**formula**	**meanings**
Triangle		$A = \dfrac{1}{2}bh$	*A* stands for area *b* stands for base *h* stands for height *l* stands for length *r* stands for radius of a circle *s* stands for side *w* stands for width π stands for the number called "pi"— equal to about $\dfrac{22}{7}$, or 3.14, or 3.1416
Square		$A = s^2$	
Rectangle		$A = lw$	
Parallelogram		$A = bh$	
Trapezoid		$A = \dfrac{1}{2}(b_1 + b_2) \times h$ (the <u>average</u> of the lengths of the two bases, times the height)	
Circle		$A = \pi r^2$	

2. PERIMETER (p) and CIRCUMFERENCE (c)

shape	diagram	formula	meanings
Square		$p = 4s$	c stands for circumference
Rectangle		$p = 2l + 2w$	d stands for diameter of a circle
Any polygon (including square and rectangle)		p = sum of the lengths of the sides	l stands for length p stands for perimeter r stands for radius of a circle s stands for side w stands for width
Circle		$c = \pi d$ or $c = 2\pi r$	π stands for the number called "pi"— equal to about $\frac{22}{7}$, or 3.14, or 3.1416

3. VOLUME (V)

shape	diagram	formula	meanings
Rectangular solid		$V = lwh$ or $V = Bh$	B stands for the area of the base of a solid (shown shaded in the drawings)
Pyramid		$V = \frac{1}{3}Bh$	h stands for height l stands for length r stands for radius of a circle or a sphere
Cylinder		$V = \pi r^2 h$ or $V = Bh$	V stands for volume w stands for width π stands for the number called "pi"—
Cone		$V = \frac{1}{3}\pi r^2 h$ or $V = \frac{1}{3}Bh$	equal to about $\frac{22}{7}$, or 3.14, or 3.1416
Sphere		$V = \frac{4}{3}\pi r^3$	

✏️ Exercises

Put a check mark inside the correct figure for each of these shapes.

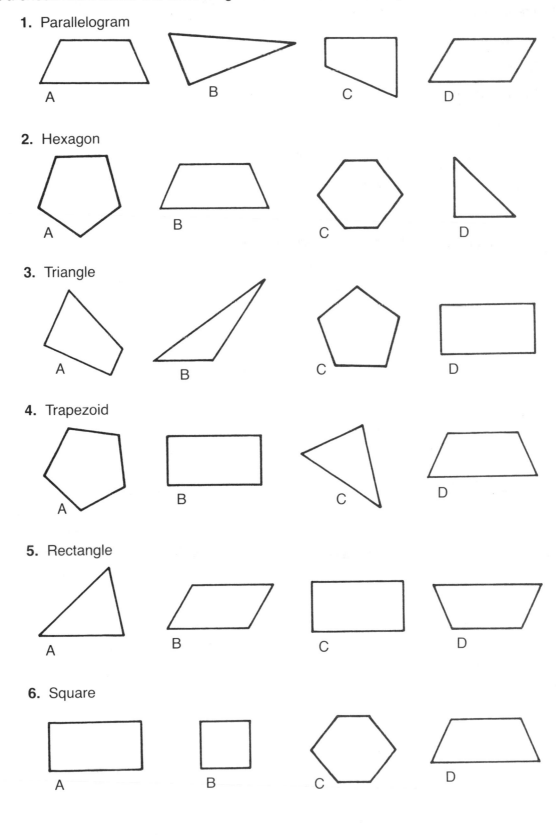

1. Parallelogram

A   B   C   D

2. Hexagon

A   B   C   D

3. Triangle

A   B   C   D

4. Trapezoid

A   B   C   D

5. Rectangle

A   B   C   D

6. Square

A   B   C   D

Write the formula for the <u>area</u> or each of the following shapes.

     7. Square _____ (use *s*)

     8. Trapezoid _____

     9. Rectangle _____

    10. Circle _____

    11. Parallelogram _____

    12. Triangle _____

Write a formula for the <u>perimeter</u> for each of the following:

    13. Square _____ (use *s*)

    14. Rectangle _____

    15. Write a formula for the <u>circumference</u> of a circle. _____

11.2 Perimeter, Area, and Volume: Examples

In this section you will find two examples completely worked out—one on area and one on volume. The solutions for both examples follow the five-step strategy of Chapter 1.

These are the five steps that make problem solving easier for you:

Step 1. WHAT?
Step 2. HOW?
Step 3. HELP
Step 4. COMPUTE
Step 5. CHECK

Here is the memory gem of Chapter 1:

WHAT! HOWARD HELPS COMPUTE CHECKS?

Each word in the sentence contains the name of one of the steps. Try to remember this sentence, and use the five steps to solve problems.

Study each step of the two examples below to see how it contributes to finding the answer.

Don't forget: In the HOW step of the strategy you have to find the number of computation steps.

And keep in mind:

Area is expressed in **square** units such as square feet or square centimeters. We use **sq** as the abbreviation for square.

Volume is expressed as **cubic** units such as cubic feet or cubic centimeters. We use **cu** as the abbreviation for cubic.

EXAMPLE 1   The XYZ Corporation has recently purchased a plot of land on which it expects to build a building. A diagram of the plot of land is shown below. What is the area of this plot (all angles are right angles)?

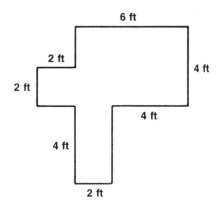

SOLUTION

    Step 1. WHAT?   To find the area of the region

Step 2. HOW? Divide the region into 3 rectangles. Use the formula for the areas of the 3 rectangles. Then add the areas. There are 4 computation steps.

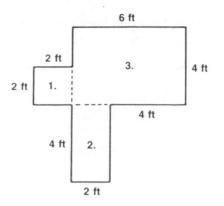

Step 3. HELP Use the formula: A = lw

Step 4. COMPUTE Region 1: A = (2)(2) = 4
Region 2: A = (4)(2) = 8
Region 3: A = (6)(4) = 24
Add: 36

The anwer is 36 sq ft.

Step 5. CHECK Ask: Is the answer reasonable?

- -

EXAMPLE 2   A shipping box used to ship the new microcomputers from the Electronic Chip Company is 2.5 ft long by 1.5 ft wide by 2 ft high. Find the volume of this box.

SOLUTION

Step 1. WHAT? To find the volume of the box

Step 2. HOW? Use the formula for the volume of a box. There is one computation step.

Step 3. HELP Use the formula: V = lwh

Step 4. COMPUTE V = (2.5)(1.5)(2) = 7.5 cu ft

Step 5. CHECK Ask: Is the answer reasonable?
Multiply in a different order:
V = (2)(1.5)(2.5) = 7.5 cu ft

Study these two examples. They help prepare you for the next section. In the next section, you will practice solving problems with perimeter, area and volume.

Many geometry problems are not like the usual word problems that you have seen throughout this book. You simply will see a figure with its dimensions and you will use a formula to find the answer.

The exercises below review basic concepts of perimeter, area and volume.

✏ Exercises

Answer the following questions. Refer back to the beginning of this chapter for any formulas that you aren't sure of. (Use $\pi = 3.14$)

1. Find the area of this triangle.

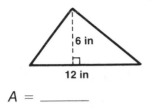

$A =$ _____

2. Find the area of this square.

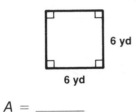

$A =$ _____

3. Find the area of this trapezoid.

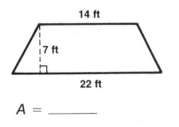

$A =$ _____

4. Find the area of this rectangle.

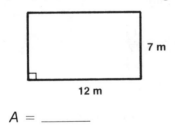

$A =$ _____

5. Find the area of this circle.

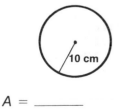

$A =$ _____

6. Find the area of this parallelogram.

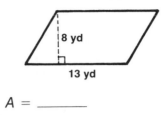

$A =$ _____

7. Find the area of this triangle.

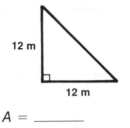

$A =$ _____

8. Find the perimeter of this square.

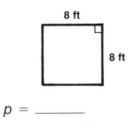

$p =$ _____

9. Find the perimeter of this pentagon.

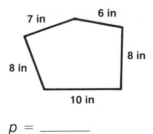

$p =$ _____

10. Find the perimeter of this rectangle.

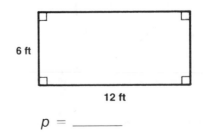

$p =$ _____

11. Find the circumference of this circle.

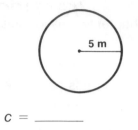

$c =$ _____

12. Find the circumference of this circle.

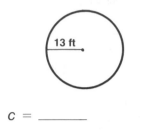

$c =$ _____

13. Find the volume of this cone.

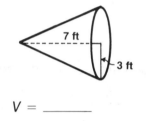

$V =$ _____

14. Find the volume of this cylinder.

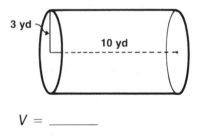

$V =$ _____

15. Find the volume of this box.

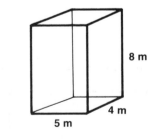

$V =$ _____

16. Find the volume of this sphere.

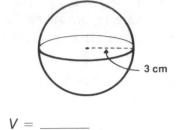

$V =$ _____

17. Find the volume of this pyramid.

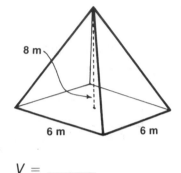

$V =$ _____

18. In which step of the 5-step strategy do we ask: is the answer reasonable?

    a. WHAT?
    b. HOW?
    c. HELP
    d. CHECK

19. In which step do we read the problem carefully?

    a. WHAT?
    b. HOW?
    c. HELP
    d. CHECK

20. In which step do we find the answer?

    a. WHAT?
    b. HOW?
    c. HELP
    d. COMPUTE

21. In which step do we figure out which operation to use?

    a. WHAT?
    b. HOW?
    c. HELP
    d. CHECK

11.3 Perimeter, Area, and Volume: Practice

In this section you will find 10 practice problems. Make sure you use the correct formula for each of the problems. Then make the calculations to get the answer. Check your answer by going over the calculations.

1. Lori is planning a small garden on the roof of her apartment house. The diagram shows the area that is available to her. Find its area.

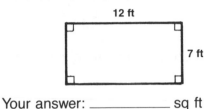

12 ft

7 ft

Your answer: _____ sq ft

2. The funnel that Patrick uses is in the shape of a cone. Find the volume of this cone. Use the formula $V = \frac{1}{3}\pi r^2 h$. ($\pi = 3.14$).

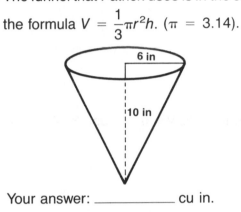

6 in

10 in

Your answer: _____ cu in.

3. Stan is a gardener. He has a large hoop that he uses to measure off round areas for flower beds. See the diagram below. Find the area of the hoop. Use the formula $A = \pi r^2$. ($\pi = 3.14$).

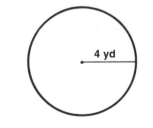

4 yd

Your answer: _____ sq yd

4. What is the area of this triangle? Use the formula $A = \frac{1}{2}bh$.

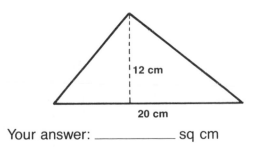

12 cm

20 cm

Your answer: _____ sq cm

5. The area of Jack's room is 120 square ft. If the length of the rectangle is 12 ft, what is the width?

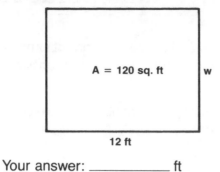

Your answer: _____ ft

6. Jim swims in a circular pool. See the diagram. Find the circumference of the pool. (Use 3.14 for π.)

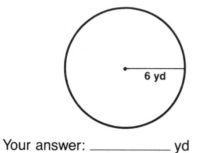

Your answer: _____ yd

7. Find the volume of this cylinder. Use the formula $V=\pi r^2 h$. (Use 3.14 for π).

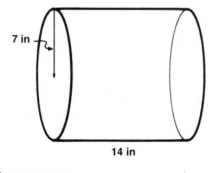

Your answer: _____ cu in.

8. Find the perimeter of this region.

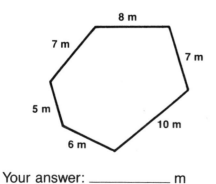

Your answer: _____ m

9. Find the volume of this box.

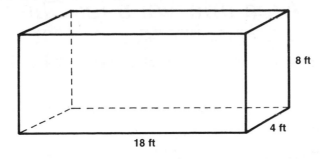

Your answer: _____ cu ft

10. Find the area of this trapezoid. Use the formula $A=\frac{1}{2}h(b_1+b_2)$

Your answer: _____ sq yd

11.4 Perimeter, Area and Volume: Checkout

The purpose of this section is to give you a chance to check out your skill in solving perimeter, area, and volume problems.

For each example, you will find four answers. Select the correct answer to the problem by circling the letter next to the answer.

For some of the problems, one of the choices is NOT GIVEN. This choice means that the correct answer is not given among the other three choices. If you can't find the correct answer, then you should circle the letter next to NOT GIVEN.

Formulas to help solve these problems can be found in Section 11.1. Use 3.14 for π throughout.

1. Here is a diagram of a corner of a park. What is the area of this triangle?

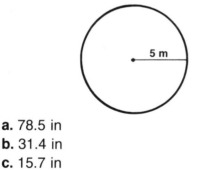

a. 48 sq ft
b. 24 sq ft
c. 14 sq ft
d. NOT GIVEN

2. What is the circumference of this circle?

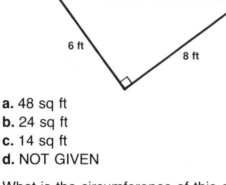

a. 78.5 in
b. 31.4 in
c. 15.7 in
d. 10.0 in

3. What is the perimeter of this region?

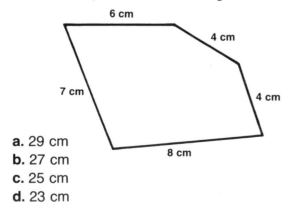

a. 29 cm
b. 27 cm
c. 25 cm
d. 23 cm

128

4. Here is a diagram of a small can of juice. What is the volume of this cylinder?

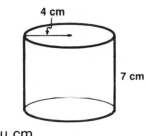

4 cm

7 cm

    a. 615.44 cu cm
    b. 351.68 cu cm
    c. 87.92 cu cm
    d. NOT GIVEN

5. Here is a diagram of the birthday gift for Ian. What is the volume of the box?

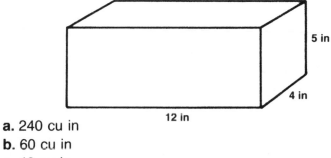

5 in

4 in

12 in

    a. 240 cu in
    b. 60 cu in
    c. 48 cu in
    d. 21 cu in

6. What is the area of this trapezoid?

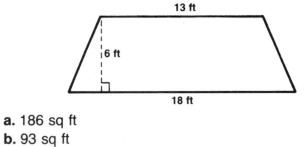

13 ft

6 ft

18 ft

    a. 186 sq ft
    b. 93 sq ft
    c. 37 sq ft
    d. NOT GIVEN

7. What is the volume of this sphere?

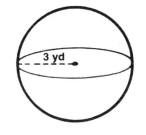

3 yd

    a. 12.56 cu yd
    b. 37.68 cu yd
    c. 113.04 cu yd
    d. NOT GIVEN

8. What is the area of this triangle?

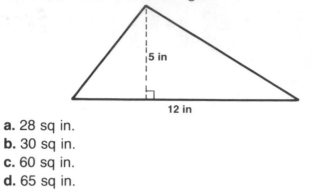

5 in

12 in

   a. 28 sq in.
   b. 30 sq in.
   c. 60 sq in.
   d. 65 sq in.

9. What is the area of this parallelogram?

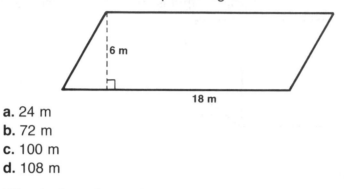

6 m

18 m

   a. 24 m
   b. 72 m
   c. 100 m
   d. 108 m

10. What is the volume of this cone?

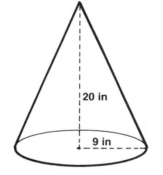

20 in

9 in

   a. 5086.9 cu in.
   b. 1695.6 cu in.
   c. 565.2 cu in.
   d. 188.4 cu in.

12. *Graphs and Statistics*

12.1 Graphs and Statistics: Getting Started

This chapter contains problems on graphs and statistics. These are the kinds of skills you will be using in this chapter:

1. Reading graphs

2. Interpreting graphs

3. Finding the mean (average), median, and mode of a set of data

In case you have forgotten these skills, the exercises below will help remind you. If you need more help, review these topics further before you start Section 12.2.

✎ *Exercises*

There are four kinds of basic graphs that you should be familiar with:

- bar graphs

- circle graphs (also called pie graphs)

- pictographs

- line graphs

In each of the following exercises, identify the type of graph.

1.

Population of the Six Largest Cities

Shanghai	👤👤👤👤👤👤
Mexico City	👤👤👤👤👤
Calcutta	👤👤👤👤👤
Seoul	👤👤👤👤
Tokyo	👤👤👤👤
Moscow	👤👤👤👤

👤 = 2 million people
👤 = 1 million people

Type of graph: _____

2. Favorite Type of Movies
   9th Graders

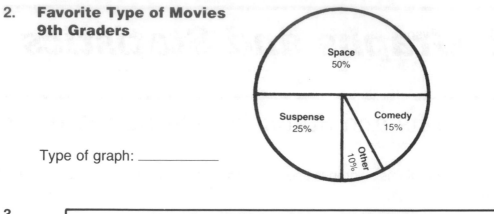

Type of graph: _____

3.

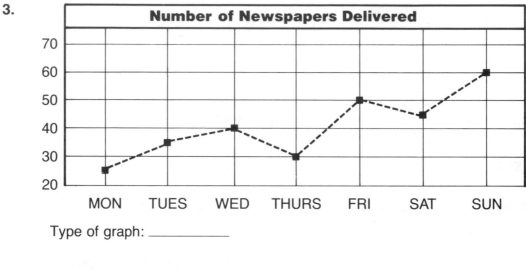

Type of graph: _____

4.

Type of graph: _____

In the next four exercises, the questions are about the graphs of the first four exercises.

5. According to the graph in Question 1, about how many people live in Mexico City?
_____ people

6. According to the graph in Question 2, what percentage of students chose suspense movies as their favorite? _____ students

7. According to the graph in Question 3, how many newspapers are delivered on Friday?
_____ newspapers

8. According to the graph in Question 4, how many people like strawberry ice cream?
_____ people

- -

The **mean** is the same as the **average**. Find the mean of a set of numbers by adding them and dividing by the number of numbers in the set. For example, the mean of 28, 47 and 24 is 33 $(28 + 47 + 24 = 99; 99 \div 3 = 33)$.

9. Find the mean of these scores: 34, 76, 53 and 17. *(Add; divide.)* _____

10. Find the mean of 124, 38, 75, 45 and 101. _____

- -

The **median** is the middle score of a set of scores. Find the median by placing the scores in order, and then finding the middle score. For example, the median of 12, 6, 13, 7 and 9 is 9.

11. What is the median of these scores: 65, 32, 12, 76, 45, 98 and 78? _____

12. What is the median of 101, 45, 125, 75, 452, 6 and 54? _____

If there is an even number of scores, there is no middle score. To find the median of an even number of scores, take the <u>middle</u> pair and compute the average of these two.

For example, here's how to find the median of 5, 2, 7, 3, 16, and 9: Order the numbers: 2, 3, 5, 7, 9, and 16. Find the middle pair:

$$2, \quad 3, \quad \mathbf{5,} \quad \mathbf{7,} \quad 9, \quad 16$$
$$\underset{\substack{middle \\ pair}}{}$$

Take the average of the middle pair:

$$\frac{5 + 7}{2} = 6 \text{ (median)}$$

13. Find the median of these scores: 84, 75, 97, 61, 80, and 92.

14. Find the median of 18, 5, 9, 2, 76 and 13.

- -

The **mode** of a set of scores is the score that appears most often. The mode of 78, 76, 68, 78, 92, 87 and 65 is 78.

15. Find the mode of 3, 3, 5, 7, 5, 8, 5, and 10.

16. Find the mode of 14, 56, 78, 2, 78, 100, 2, and 78.

In this section you will find two examples completely worked out. The solutions for both examples use the five-step strategy of Chapter 1.

These are the five steps that make problem solving easier for you:

> **Step 1.** WHAT?
> **Step 2.** HOW?
> **Step 3.** HELP
> **Step 4.** COMPUTE
> **Step 5.** CHECK

In Chapter 1 we introduced a memory gem to help you remember the five steps:

WHAT! HOWARD HELPS COMPUTE CHECKS?

Each word in this sentence contains the name of one of the steps. Try to remember this sentence, and use the five steps to solve problems.

Study each step in the two examples below to see how it contributes to finding the answer.

Don't forget: In the HOW step of the strategy you have to answer the question:

How many computation steps will it take to solve the problem?

Many problems require only one computation step. These are called 1-step problems. Other problems require more than one computation step. These are called 2-step, 3-step, etc. depending upon the number of computation steps it takes. Example 1 is a 3-step problem, while Example 2 is a 2-step problem.

- -

EXAMPLE 1   Aliza made a graph of the number of games her high school soccer team won every year. This is what the graph looks like. How many more games did the soccer team win in 1986 than in 1979?

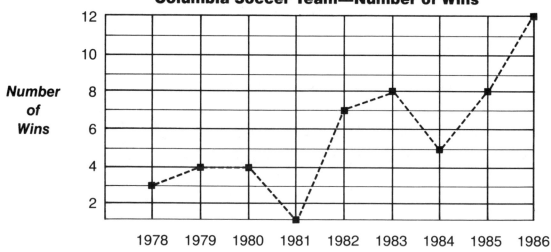

SOLUTION

Step 1. WHAT?   How many more wins did the soccer team get in 1986 than in 1979?

Step 2. HOW?   1) Find how many wins in 1986.

    2) Find how many wins in 1979.

    3) Subtract 2) from 1). There actually is only one computation step.

Step 3. HELP $12 - 4 = ?$

Step 4. COMPUTE $12 - 4 = 8$

    The answer is 8 wins.

Step 5. CHECK   Check the number of wins in 1979 and 1986

    Add: $8 + 4 = 12$

- -

EXAMPLE 2   Here are Ian's spelling test scores for the past 6 weeks: 67, 87, 86, 90, 56, 77 and 72. What is the median?

SOLUTION

Step 1. WHAT?   What is the median of the spelling scores?

Step 2. HOW?   1) Put the scores in order

    2) Find the middle score. There is no computation necessary.

Step 3. HELP   Here are the scores in order:

    56, 67, 72, 77, 86, 87, 90

Step 4. COMPUTE The median (middle score) is 77.

Step 5. CHECK   Look at the numbers again:

    —Are they in order?

    —Do you have the middle score?

- -

EXAMPLE 3   The mean of a set of six test scores is 7. The scores are: 7, 9, 6, 8, 4, and one missing number. What is the missing score?

$$\frac{7+9+6+4+\boxed{7}+11}{2} = 7$$

SOLUTION

Step 1. WHAT?   What is the missing score?

Step 2. HOW?   1) Find the total of the scores by multiplying the average score (7) by the number of scores (6).

    2) Find the total of the scores we know by adding.

    3) Subtract 2) from 1).

    There are 3 computation steps.

Step 3. HELP   Here are the setups:

Step 1	Step 2	Step 3
$7 \times 6 = ?$	$7 + 9 + 6 + 8 + 4 = ?$	Subtract the answer to Step 2 from the answer to Step 1.

Step 4. COMPUTE

Step 1	Step 2	Step 3
$7 \times 6 = 42$	$7 + 9 + 6 + 8 + 4 = 34$	$42 - 34 = 8$

The missing score is 8.

Step 5. CHECK   Ask: Is this answer reasonable?
To check: Add up all six scores, including your answer, and divide by 6 to find the average score.
$7 + 9 + 6 + 8 + 4 + 8 = 42$
$42 \div 6 = 7$

Study all three of these examples. Pay particular attention to Example 3. It's tricky, and it turns up a lot on tests.

These three examples will help prepare you for the next section. In the next section, you will practice solving word problems with graphs and statistics. The exercises below review some additional basic concepts about graphs and statistics in preparation for the next section.

✎ Exercises

1. Find the mean (average) of: 18, 56, 30, 24, 14, and 8. _____

2. Find the mode of: 23, 87, 62, 34, 87, 63, and 7. _____

3. Find the median of: 45, 11, 61, 87, 99, and 6. _____

4. Find the mean (average), median, and mode of these scores: 23, 74, 38, 102, 33, 23 and 8.

   a. mean _____
   b. median _____
   c. mode _____

5. The average of a set of five scores is 6. The scores are 3, 9, 7, 6, and one other number. What is the missing score? _____

6. The mode of seven scores is 46. The scores consist of 4, 23, 87, 90, 46, 70 and one missing score. What is the missing score? _____

7. The median of seven scores is 40. The scores are 23, 46, 77, 80, 17, 37 and one missing number. What is the missing score? _____

Use this graph to answer Questions 8–10.

THE TIGERS BASEBALL TEAM

Number of Wins by Month

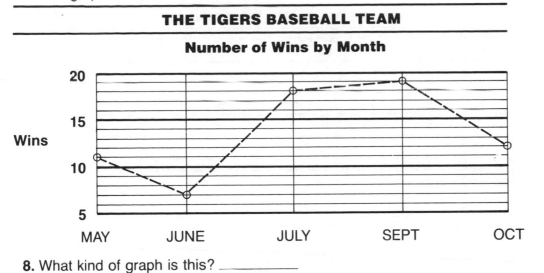

8. What kind of graph is this? _____

9. In which month did the Tigers win the fewest games? How many did they win that month? _____

10 In which month did the Tigers win the most games? How many did they win in that month? _____

Use this graph to answer Questions 11–14.

How Marge spends her salary

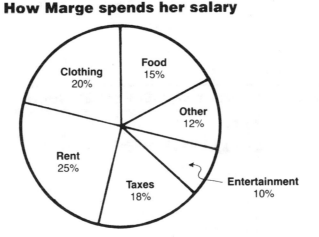

11. What kind of graph is this? _____

12. What percent of her salary does Marge spend on clothing? _____

13. What percent of her salary does Marge spend on food? _____

14. Marge earns $800 a month. How much does she spend on rent each month? _____

Use this graph to answer Questions 15–18.

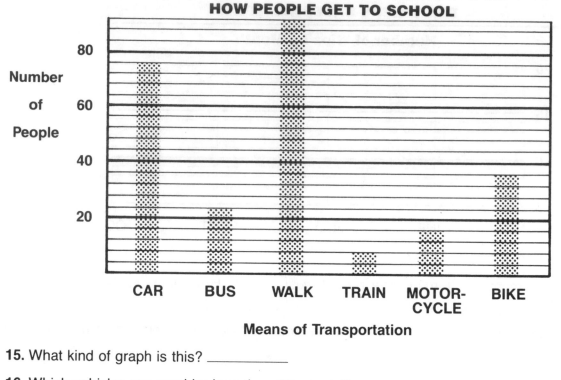

HOW PEOPLE GET TO SCHOOL

Means of Transportation

15. What kind of graph is this? _____

16. Which vehicles are used by less than 20 people? _____

17. How many people walk to school? _____

18. How many people use the power of their own legs to get to school? (Walk & bike)
_____

Use this graph to answer Questions 19–22.

Favorite Sport of High School Students	
Volleyball	👤👤👤
Basketball	👤👤👤👤👤
Football	👤👤👤
Baseball	👤👤👤
Soccer	👤👤👤👤👤
Tennis	👤👤👤👤

👤 = 10 students
👤 = 5 students

19. A survey was taken at Webster High School to find out the favorite sport among students. The results of the survey are shown in the graph above. What kind of graph is this? _____

20. How many students chose baseball as their favorite sport? _____

21. How many people selected soccer as their favorite sport? _____

22. How many students preferred a game in which it is legal to kick the ball? _____

The next questions are about the 5-step procedure for solving problems.

23. In which step of the 5-step procedure do we ask: is the answer reasonable?

    a. WHAT?
    b. HOW?
    c. HELP
    d. CHECK

24. In which step do we find the answer?

    a. WHAT?
    b. HOW?
    c. HELP
    d. COMPUTE

25. In which step do we find out which operation to use?

    a. WHAT?
    b. HOW?
    c. HELP
    d. CHECK

26. In which step do you read the problem carefully to find what the question wants you to do?

    a. WHAT?
    b. HOW?
    c. HELP
    d. COMPUTE

12.3 Graphs and Statistics: Practice

In this section you will find 10 problems to practice. When possible, use the 5-step procedure to solve these problems, and make sure to check each answer.

There are four graphs used in these problems. These graphs will be used again in Section 10.4.

Use the graph below to answer Questions 1–3.

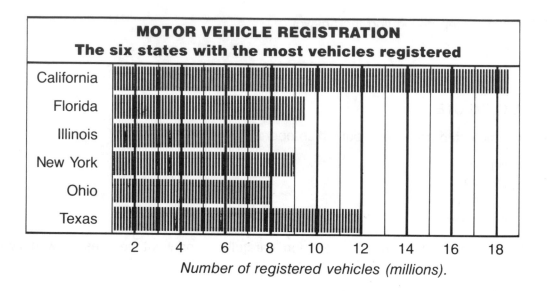

MOTOR VEHICLE REGISTRATION
The six states with the most vehicles registered

Number of registered vehicles (millions).

1. This graph shows the six states with the most registered motor vehicles. About how many vehicles are registered in Illinois?
   Your answer: _____ vehicles

2. About how many more vehicles are registered in Texas than in Ohio?
   Your answer: _____ vehicles

3. How many vehicles are registered in New York and Florida together?
   Your answer: _____ vehicles

Use the graph below to answer Questions 4–5.

Results of Class Election for President

Kristina	★ ★ ★
Tina	★ ★ ★ ★
Kevin	★ ★ ★ ★ ★ ☆
Todd	★ ★ ★ ☆
Lesley	★ ★ ★ ★ ★ ★
Rachel	★ ★

★ = 10 votes
☆ = 5 votes

4. This graph shows the results of a class election for president. Who got the most votes? How many votes did the winner get?

Your answers: _____ , _____ votes

5. How many more votes did Tina get than Rachel?

Your answer: _____ votes

Use the graph below to answer Questions 6–8.

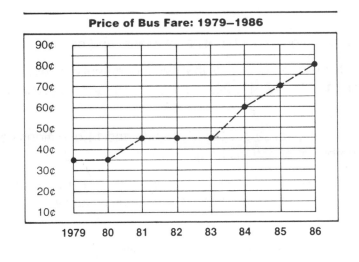

Price of Bus Fare: 1979–1986

6. This graph shows the bus fare on Jan.1 from 1979 to 1986. In which January did the biggest jump in bus fare take place? How much did it increase that year?
Your answer: _____ , _____ cents

7. How much more was the bus fare in 1984 than it was in 1979?
Your answer: _____ cents

8. There was no rise in the bus fare in three different years. What were the three years?
Your answer: _____ , _____ , _____

Use the graph below to answer Questions 9–10.

How Jeff Spends His Free Time

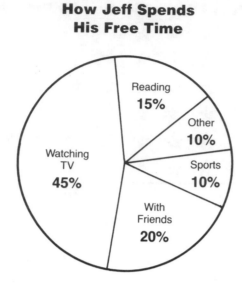

9. This graph shows how Jeff spends his free time. Suppose Jeff has 30 hours of free time each week. About how many hours does he spend playing sports?
   Your answer: _____ hours

10. Suppose Jeff has 40 hours of free time. About how many hours does he spend with friends?
    Your answer: _____ hours

Answer the following questions about means, medians, and modes:

11. Renée counts the number of calls that her department gets during a 10-day period. The numbers are 17, 19, 12, 17, 28, 22, 17, 12, 18, and 13. What is the mode of these numbers? _____

12. In a series of tests, Jesse got 16, 12, 18, 20, and 14 questions correct. What was the average number of questions he got correct? _____

13. Family incomes of the 6 families that live on Park Place are: $18,000; $37,000; $22,000; $28,000; $19,000. What is the median family income of these families?
    _____

14. The mean of a set of data is 34. The data consist of six numbers: 72, 21, 12, 65, 14, and one missing number. Find the missing number _____

12.4 Graphs and Statistics: Checkout

The purpose of this section is to give you a chance to check your skills in solving problems with graphs and statistics.

For each example, you will find four answers. Select the correct answer to the problem by circling the letter next to the answer.

For some of the problems, one of the choices is NOT GIVEN. This choice means that the correct answer is not given among the other three choices. If you can't find the correct answer, then circle the letter next to NOT GIVEN.

The first 12 problems use the same four graphs as in Section 10.3.

Use the graph below to answer Questions 1-3.

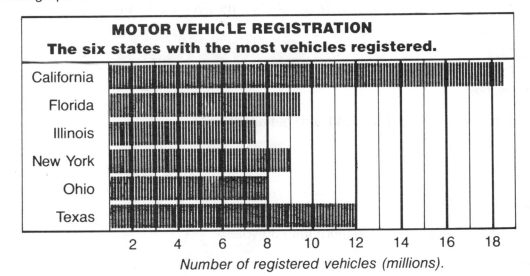

1. Look at the bar graph above. About how many vehicles are registered in the states of Florida and Illinois together?

   a. 18 million
   b. 15 million
   c. 9.2 million
   d. NOT GIVEN

2. About how many more registered vehicles does California have than Ohio?

   a. 8.5 million
   b. 9.5 million
   c. 10.5 million
   d. NOT GIVEN

3. Which state has about 12 million registered vehicles?

   a. Florida
   b. New York
   c. Ohio
   d. Texas

143

Use the graph below to answer Questions 4-6.

Results of Class Election for President

Kristina	★ ★ ★
Tina	★ ★ ★ ★
Kevin	★ ★ ★ ★ ★ ☆
Todd	★ ★ ★ ☆
Lesley	★ ★ ★ ★ ★ ★ ★
Rachel	★ ★

★ = 10 votes
☆ = 5 votes

4. Look at the pictograph above. Which student received 35 votes?

   a. Kristina
   b. Tina
   c. Todd
   d. Rachel

5. How many votes did Tina and Kevin get altogether?

   a. 95
   b. 105
   c. 190
   d. NOT GIVEN

6. How many people received more than 30 votes?

   a. 3
   b. 4
   c. 5
   d. 6

Use the graph below to answer Questons 7-9.

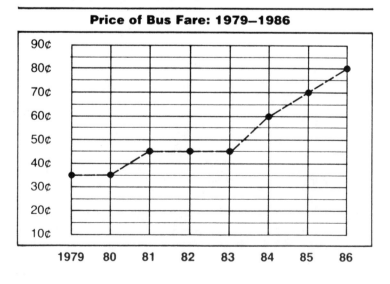

Price of Bus Fare: 1979–1986

7. Look at the line graph on the previous page. How much more was the bus fare in 1986 than it was in 1979?

    a. 35 cents
    b. 40 cents
    c. 45 cents
    d. NOT GIVEN

8. In which year was the bus fare 60 cents?

    a. 1983
    b. 1984
    c. 1985
    d. 1986

9. What was the amount of increase in the bus fare from 1979 to 1980?

    a. 5 cents
    b. 10 cents
    c. 15 cents
    d. NOT GIVEN

Use the graph below to answer Questions 10-12.

How Jeff Spends His Free Time

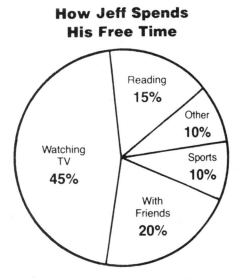

10. Look at the circle graph above. What does Jeff do during 15% of his spare time?

    a. He is with his friends.
    b. He plays sports.
    c. He reads.
    d. He watches TV.

11. Suppose Jeff has 40 hours of free time. How many hours does he spend playing sports?

    a. 10
    b. 8
    c. 4
    d. 2

12. Suppose Jeff has 20 hours of free time. How many hours does he spend playing sports and reading?

a. 5
b. 6
c. 10
d. NOT GIVEN

13. Aliza received these grades in math during the first marking period: 87, 72, 93 and 80. What is the mean of these grades?

a. 81
b. 82
c. 83
d. NOT GIVEN

14. Mike is the team's star. In the last 10 basketball games he scored 24, 18, 31, 17, 22, 19, 10, 28, 39 and 28 points. What is the median of these scores?

a. 31
b. 28
c. 23
d. 22

15. Cathy keeps careful records of her bowling games. Here is a chart of her last 10 games.

127	168
145	200
192	204
128	165
255	182

What is the median of these scores?

a. 175
b. 187
c. 198
d. NOT GIVEN

16. Stan is a waiter in a diner. He keeps records of how many people come in to eat. Here is his tally for the last 15 groups to come in:

2	4	1
4	3	2
2	2	2
1	3	3
2	4	6

What is the mode of these numbers?

a. 1
b. 2
c. 3
d. 4

17. Janet sells magazines. She made a list of the prices of the 12 most popular magazines:

$3.00	$2.50
$2.25	$1.75
$2.50	$1.50
$1.50	$3.25
$1.00	$1.75
$2.50	$3.50

What is the mode of these prices?

a. $3.00

b. $2.50

c. $1.75

d. $1.50

18. Pat looked at the bills from the last five times she went food shopping. Here are the totals: $42, $76, $38, $28 and $21. What is the mean of these bills?

a. $41

b. $38

c. $46

d. NOT GIVEN

19. Lisa looked at the last 11 electric bills for her apartment. Here are the totals:

$22	$36
$27	$42
$36	$28
$19	$33
$34	$25
$18	

What is the median of these 11 bills?

a. $22

b. $27

c. $33

d. NOT GIVEN

20. In one five-day week, Phil averaged $19 a day in tips for waiting on tables at the local coffee shop. On the first four days of the week, Phil made $17, $13, $19, and $22. How much did he make on the fifth day?

a. $17

b. $19

c. $24

d. NOT GIVEN

13. *REVIEW TEST C*

This test covers the topics of Chapters 10, 11 and 12. It consists of 20 questions covering all three topics—1) measurement, 2) perimeter, area and volume and 3) graphs and statistics.

There are four answers for each example. Select the correct answer to the problem by circling the letter next to the answer.

For some of the problems, one of the choices is NOT GIVEN. This choice means that the correct answer is not given among the other three choices. If you can't find the correct answer, then circle the letter next to NOT GIVEN.

- -

Use the bar graph below to answer Questions 2 and 3. The graph shows the results of a survey. It shows the types of music people like.

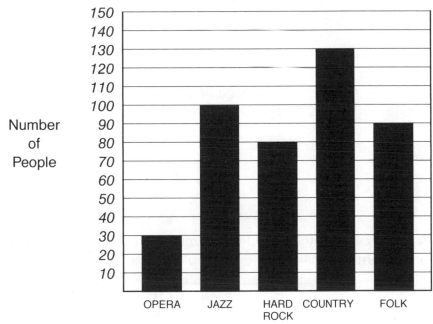

TYPES OF MUSIC PEOPLE LIKE

1. How many people like opera?

   a. 60
   b. 50
   c. 40
   d. 30

2. How many more people like jazz than hard rock?

   a. 30
   b. 20
   c. 10
   d. NOT GIVEN

3. Sherrill wants to put a fence around his garden. The garden is a rectangle 18 ft by 7 ft. How long will the fence be if the entire garden will be enclosed?

a. 126 ft
b. 60 ft
c. 50 ft
d. NOT GIVEN

4. Mike's goal is to run 3 miles, which is 10 times around the school track. How many feet is once around the track?

a. 1760
b. 5280
c. 15,840
d. NOT GIVEN

5. Jack's golfing scores for the summer were: 92, 87, 98, 85, 90, 88, and 87. What is the median of these scores?

a. 85
b. 87
c. 88
d. 90

6. Steve counted the number of calories in the food he ate for the week. Here are the 7-day totals: 2300, 2600, 2900, 3000, 2000, 2700 and 4100. What is the average number of calories per day?

a. 2100
b. 2300
c. 2600
d. NOT GIVEN

7. The land that the Smythes own is in the form of a trapezoid (see diagram below). They bought this land for $50 per square foot. What was the cost of the land?

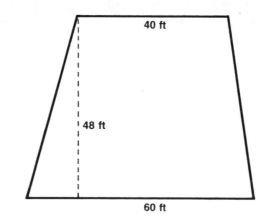

40 ft

48 ft

60 ft

a. $60,000
b. $80,000
c. $120,000
d. $160,000

Use the line graph below to answer Questions 8 and 9. The graph shows the number of people who tried out for the soccer teams at school.

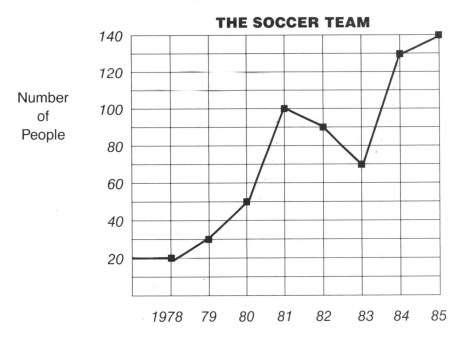

THE SOCCER TEAM

Number of People

140
120
100
80
60
40
20

1978 79 80 81 82 83 84 85

8. How many more people came out for soccer in 1985 than in 1979?

    a. 90
    b. 100
    c. 110
    d. NOT GIVEN

9. In which year was there the largest increase in the number of people trying out for soccer?

    a. 1980
    b. 1981
    c. 1983
    d. 1984

- -

10. The scout group has marched 4000 m so far in the parade. Their leader, Derek, says that they have four times further to go. What is the total distance of the march?

    a. 16 km
    b. 20 km
    c. 160 km
    d. 200 km

11. Aliza sent four packages to her friend in England. These are the weights of the four packages: 12 oz, 15 oz, 1 lb 3 oz and 1 lb 6 oz. How much do these packages weigh altogether?

   a. 3 lb 4 oz
   b. 4 lb
   c. 4 lb 4 oz
   d. 4 lb 10 oz

12. The length of a room measures 451 cm. What is an equivalent way to state this length?

   a. 4510 m
   b. 45 m 1 cm
   c. 4 m 51 cm
   d. NOT GIVEN

13. What is the volume of the can in the diagram below? (Use 3.14 for π.)

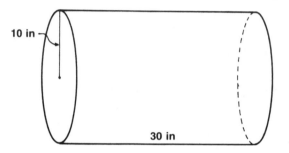

   a. 9420 cu in.
   b. 9320 cu in.
   c. 942 cu in.
   d. NOT GIVEN

14. A large mainframe computer prints out about 40 yards of paper per hour. How many feet is this?

   a. 400
   b. 240
   c. 120
   d. 60

15. Jan was preparing for a party. She made punch from three juices: apple juice—500 ml; cranberry juice—750 ml; and pear juice—400 ml. She poured all three juices into a 3-liter container. How much capacity was left over after she poured the three juices into the container?

   a. 650 ml
   b. 750 ml
   c. 1650 ml
   d. NOT GIVEN

16. Rex watched TV from 6:00 p.m. to 10 p.m. He watched an average of 14 commercials per hour. He made an exact record of the number of different commercials each hour. Here is his record:

Time	Number of commercials
6–7 p.m.	12
7–8 p.m.	13
8–9 p.m.	?
9–10 p.m.	16

What is the number of commercials from 8–9 p.m.?

a. 12
b. 13
c. 14
d. 15

- -

Use the pictograph below to answer questions 17 and 18. The graph shows the results of a recent survey about technology.

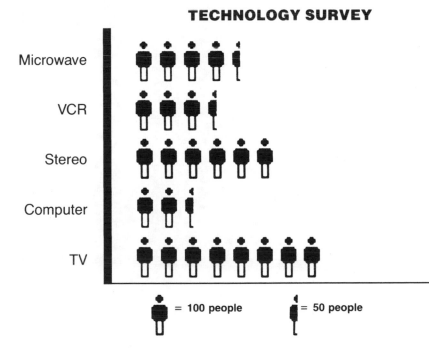

17. How many people own a computer?

    a. 200
    b. 250
    c. 300
    d. 350

18. How many more people own a microwave than own a computer?

    a. 100
    b. 150
    c. 200
    d. 250

19. A table is 3 m long and 250 cm wide. What is the area of the table?

    a. 5.5 sq m
    b. 6 sq m
    c. 8.5 sq m
    d. NOT GIVEN

20. Gloria works at the local coffee shop. Her tips for the last 7 days have been: $34, $31, $40, $19, $31, $23 and $31. What is the mode of these tips?

    a. $26
    b. $29
    c. $30
    d. NOT GIVEN

14. *FINAL EXAM*

This exam covers all nine topics of the book. There are 25 problems to solve. Make sure you use the 5-step approach to help you get the answer and to check the answer.

There are four answers for each example. Select the correct answer to the problem by circling the letter next to the answer.

For some of the problems, one of the choices is NOT GIVEN. This choice means that the correct answer is not given among the other three choices. If you can't find the correct answer, then circle the letter next to NOT GIVEN.

1. George spent $\frac{1}{6}$ of his weekly salary on a new camera. His salary each week is $654. How much did he pay for the camera?

    a. $545
    b. $190
    c. $109
    d. NOT GIVEN

2. What is the volume of this box?

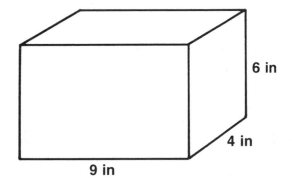

    a. 18 cu in.
    b. 36 cu in.
    c. 54 cu in.
    d. 216 cu in.

3. Fred says that the ratio of good ideas to bad ideas is 3 to 10. If there are 700 bad ideas, how many good ideas are there?

    a. 210
    b. 200
    c. 140
    d. NOT GIVEN

4. Brad vaulted 3.3 m. His friend Kate vaulted 1.3 times Brad's height. How high did Kate vault?

   a. 4.29 m
   b. 4.6 m
   c. 5.129 m
   d. NOT GIVEN

5. Jan's plane arrived right on schedule at 11:05 a.m. after a flight of 8 hours 37 minutes. What time did she leave?

   a. 2:28 a.m.
   b. 2:42 a.m.
   c. 3:38
   d. NOT GIVEN

6. The label says that the food weighs 4 kg 236 g. How many grams is this?

   a. 636 g
   b. 1636 g
   c. 4236 g
   d. NOT GIVEN

7. It was quite a show! Around 80,000 saw the game at the stadium. Another 120,000,000 people watched the game on TV. How many more people saw the game on TV than were at the game?

   a. 119,910,000
   b. 119,920,000
   c. 120,020,000
   d. 120,080,000

- -

Use the graph below to answer questions 8 and 9. The graph shows Sheila's budget for the year.

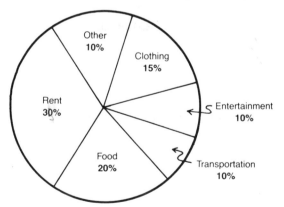

8. What percent of her budget has Sheila set aside for rent?

   a. 10%
   b. 20%
   c. 30%
   d. 40%

9. How much more does Sheila plan on spending for clothing than for entertainment?

    a. 5%
    b. 10%
    c. 15%
    d. 20%

10. The price for one share of stock of the Focus Computer Co. was $6\frac{5}{8}$ in the morning. In the evening one share was selling at $7\frac{1}{4}$. How much higher was the share selling for in the evening than in the morning?

    a. $1\frac{5}{8}$
    b. $1\frac{1}{4}$
    c. $\frac{5}{8}$
    d. NOT GIVEN

11. Mike and Vanessa made videos for their class projects. Mike's tape ran for 17 minutes; Vanessa's tape ran for 22 minutes. How much longer is Vanessa's tape—in seconds?

    a. 600
    b. 500
    c. 300
    d. 200

12. Here are the number of points that Carla scored in the last six basketball games: 16, 12, 23, 24, 19 and 14. What is the average number of points that Carla scored in these last six games?

    a. 12
    b. 14
    c. 16
    d. 18

13. The software program that Jim needed to figure out his taxes was on sale for 25% off. The list price was $236. How much was the program selling for?

    a. $59
    b. $177
    c. $187
    d. NOT GIVEN

14. Pat needed clay for the pottery that she was making. She ordered 43 boxes of clay. Each box weighs 57 lb. How much did the entire order weigh?

    a. 2351 lb
    b. 2431 lb
    c. 2451 lb
    d. 2471 lb

15. Dan took the train to Philadelphia. The train left at 8:05 a.m. and arrived at 12:34 p.m. It stopped three times for 7 minutes each. How long was the actual train ride?

   a. 4 hours 8 minutes
   b. 4 hours 29 minutes
   c. 4 hours 39 minutes
   d. 4 hours 50 minutes

16. The ratio of hard problems to easy problems is 2 to 5. If there are 12 hard problems, how many easy problems are there?

   a. 10
   b. 25
   c. 30
   d. 60

17. Harriet is a saver. She now has 1345 nickels in her collection. How much money is that?

   a. $67.05
   b. $67.20
   c. $67.30
   d. NOT GIVEN

18. Len measured the length of the computer screen. It is 22 cm long. It is $\frac{2}{3}$ the size of the screen in the main building. How long is the bigger screen?

   a. 11 cm
   b. 30 cm
   c. 33 cm
   d. NOT GIVEN

19. Keith says that the new machine can put 25 labels on envelopes every minute. How many labels can the machine do in three hours?

   a. 5000
   b. 4500
   c. 2500
   d. 1000

20. The new space miniseries "Galaxy Darkness" will be shown in five parts. Each part is two hours long and has 12 commercial breaks lasting two minutes each. How long is the actual series?

   a. 182 minutes
   b. 440 minutes
   c. 480 minutes
   d. 576 minutes

21. Gregg said that 22% of all the money he makes must be paid as taxes. He made $9000 last year. How much did he pay in taxes?

    a. $3600

    b. $2200

    c. $1960

    d. NOT GIVEN

22. Joan measured the dimensions of a playground. It is a quadrilateral with sides equal to 27 ft, 34 ft, 45 ft and 55 ft. What is the perimeter of the playground?

    a. 161 ft

    b. 162 ft

    c. 171 ft

    d. NOT GIVEN

23. What is the mode of these tolls that Charlie had to pay on his long trip: $.15, $.25, $.35, $.25, $2.00, $.40, $1.75, $.15, $.35 and $.35?

    a. $.15

    b. $.25

    c. $.35

    d. $.45

24. There is a ribbon that is perfect for decorations. It is 440 cm long and has to be cut into 24 equal pieces. How long will each piece be?

    a. 18.9 cm

    b. 18.8 cm

    c. 18.5 cm

    d. NOT GIVEN

25. The students have $1\frac{1}{2}$ hours to do the 25 questions on the test. Bill finished the exam in 54 minutes. How much time does he have left to check his answers?

    a. 96 minutes

    b. 56 minutes

    c. 46 minutes

    d. 36 minutes